Solar Powered Charging Infrastructure for Electric Vehicles

A Sustainable Development

Solar Powered Charging Infrastructure for Electric Vehicles

A Sustainable Development

Edited by
Larry E. Erickson • Jessica Robinson
Gary Brase • Jackson Cutsor

CRC Press
Taylor & Francis Group
Boca Raton London New York

CRC Press is an imprint of the
Taylor & Francis Group, an **informa** business

Cover photo credits provided by Envision Solar International, Inc. (left); Tesla Motor Inc. (upper right); and Vundelaar, Roos Korthals Altes [Fastned fast changing station] (lower right).

CRC Press
Taylor & Francis Group
6000 Broken Sound Parkway NW, Suite 300
Boca Raton, FL 33487-2742

First issued in paperback 2017

ISBN 13: 978-0-8153-8371-0 (pbk)
ISBN13: 978-1-4987-3156-0 (hbk)

Library of Congress Cataloging-in-Publication Data

Names: Erickson, L. E. (Larry Eugene), 1938- editor. | Robinson, Jessica, 1994- editor. | Brase, Gary, editor. | Cutsor, Jackson, editor.
Title: Solar powered charging infrastructure for electric vehicles : a sustainable development / editors, Lary E. Erickson, Jessica Robinson, Gary Brase, and Jackson Cutsor.
Description: Boca Raton : CRC Press, Taylor & Francis Group, [2017] | "Solar powered charging infrastructure for EVs is a rapidly evolving field. With the recent increase in the number of EVs on the roads, there is a need for a comprehensive description of the evolving charging infrastructure, particularly SPCS. The authors attempt to give readers information on the existing solar powered charging infrastructure, while discussing its advantages, mainly in light of sustainable development; air quality improvement, and reduced dependence on fossil fuels"--Provided by publisher. | Includes bibliographical references and index.
Identifiers: LCCN 2016007998 | ISBN 9781498731560 (alk. paper)
Subjects: LCSH: Battery charging stations (Electric vehicles) | Electric vehicles--Power supply. | Electric vehicles--Batteries. | Photovoltaic power generation. | Photovoltaic power systems. | Sustainable development.
Classification: LCC TK2943 .S65 2017 | DDC 388.3--dc23
LC record available at https://lccn.loc.gov/2016007998

Visit the Taylor & Francis Web site at
http://www.taylorandfrancis.com

and the CRC Press Web site at
http://www.crcpress.com

Contents

Foreword

Engineers work to develop new technologies to advance our daily lives. While some technologies make sense to the developing engineers, often economics or social impacts and acceptance create challenges for the adoption of new technologies. This book provides technical, economic, and social implication information about two technologies that have seen a diverse response related to integration and acceptance. The use of solar energy within the charging infrastructure for electric vehicles provides some key opportunities related to global usage of these vehicles as well as reduced emissions for countries struggling with air quality as industrialization and automobile numbers have increased.

This book is an excellent example of the synergies in higher education that help advance state-of-the-art technologies, educate our future engineering workforce, and disseminate challenges, issues and solutions for today's and tomorrow's energy challenges. Faculty from five different departments across Kansas State University have combined to provide their expertise in the areas of economics, psychology, electric power, air quality, and renewable energy to develop a comprehensive review of using solar power for electric vehicles. Additionally, engineering undergraduate students from across the country contributed as part of an extension of their National Science Foundation Research Experience for Undergraduate program. The book was also made possible through the support of the Black and Veatch Foundation through the "Building a World of Difference" Program.

This book will be a useful resource for a multitude of audiences, ranging from the general public, an introduction to renewables class, introduction to engineering class, or even for an upper level engineering elective. It responds directly to two of the U.S. National Academy of Engineering Grand Challenges for Engineering: (1) make solar energy economical and (2) restore and improve urban infrastructure.

I applaud the editors and contributors for developing this helpful tool to share and help advance this topic for generations to come.

Dr. Noel Schulz
IEEE Fellow
Kansas State University

Preface

Unless someone like you cares a whole awful lot, nothing is going to get better. It's not.

Dr. Seuss

Since 2009, Kansas State University has had about 10 to 18 college students who have annually participated in a 10-week summer research experience for undergraduates program, Earth, Wind, and Fire: Sustainable Energy in the 21st Century, with most of the financial support provided by the National Science Foundation. Each summer we have had a team project related to generating electricity using solar panels in parking lots. The concept of solar powered charging stations (SPCSs) for electric vehicles (EVs) grew out of the early dialog as interest and developments in EVs progressed. Shortly after publication of our second manuscript (Robinson et al., 2014) we received an invitation to write a book on SPCSs for EVs. Because of all of the different significant issues related to SPCSs and EVs, we decided to write this book. In this age of sustainable development, environmental considerations are receiving greater consideration, and we have included these topics in this book.

This book is written for all people, everywhere, because the transition to solar and wind energy for the generation of electricity and the electrification of transportation is going to impact everyone. In the next 50 years, electricity from solar energy is going to become much more important, and EVs will grow in numbers from more than one million in service now to much larger numbers. There are already many SPCSs in the world. However, the transition from the present number of parking spaces with solar panels over them to having over 200 million parking spaces with shaded parking provided by SPCSs will not be easy. It will benefit from having an educated public that understands the values, issues, and benefits of SPCSs and EVs. This book is an introduction to the topics related to SPCSs and EVs. We address the social, environmental, economic, policy, and organizational issues that are involved, as well as the complex and multidisciplinary dimensions of these topics. Related topics include infrastructure for EV charging, batteries, energy storage, smart grids, time-of-use (TOU) prices for electricity, urban air quality, business models for SPCSs, government regulation issues, taxes, financial incentives, and jobs.

Globally, the expenditures for the generation and use of electricity and for automobile travel are each more than one trillion dollars per year. The transition to more electricity from wind and solar generation with 200 million SPCSs and EVs is expensive and entails significant capital investment.

This transition has already begun, though, for several reasons. One reason is because the prices of solar panels and batteries are decreasing. Another reason is that greenhouse gas emissions are reduced by generating electricity with wind and solar energy and by electrifying transportation.

The Paris Agreement on Climate Change adopted on December 12, 2015 is a major step forward in many respects. There is now almost unanimous agreement that it would be good to reduce greenhouse gas emissions. This book addresses one way to do it. In order to accomplish the goal of achieving a balance between emissions and sinks for carbon dioxide before 2100, significant progress in transitioning to SPCSs and EVs is needed. Two of the largest sources of carbon dioxide emissions are the generation of electricity and transportation. Globally, air quality is a major issue in many large urban areas, and the transition to EVs will be very beneficial to the health for those living in these cities. The transportation sector is one of the largest causes of air pollution, and eliminating combustion emissions is a good way to improve air quality.

Regulatory and policy issues are included in the book because there are currently limitations on the sale of electricity in many locations. The financial and environmental aspects contribute to the complexity of business models that may be used to pay for and profit from constructing and operating SPCSs. Those involved in government, regulatory commissions, banking, and finance need to understand the value and importance of SPCSs for EV infrastructure. Members of environmental organizations who want to encourage environmental progress will benefit from reading this book. We hope the book will also be helpful to those interested in sustainable development and the best pathways to a sustainable world.

You as a reader can make a difference. Some readers can make a bigger difference because of their ability to influence policy or corporate decisions, but there are actions that each reader can take. Actions by everyone can add to significant change toward a more sustainable world. This is something everyone wants.

Reference

Robinson, J., G. Brase, W. Griswold, C. Jackson, and L.E. Erickson. 2014. Business models for solar powered charging stations to develop infrastructure for electric vehicles, *Sustainability* 6: 7358–7387.

Larry E. Erickson
Jessica Robinson
Gary Brase
Jackson Cutsor

Acknowledgments

Many people have been supportive and helpful in the effort to advance the science, technology, and supporting processes that are important to developing an infrastructure with many parking lots full of solar powered charging stations (SPCSs) for electric vehicles (EVs). Developing the manuscript for this book has been a team project, and we thank all who have helped. We are attempting to give appropriate credit by showing chapter authors. Gary Brase, Jackson Cutsor, Larry E. Erickson, and Jessica Robinson have helped write several chapters and edit the chapters; they are shown as editors of the book.

The National Science Foundation has provided financial support for 10 students each summer since 2009 for the Earth, Wind, and Fire: Sustainable Energy in the 21st Century Research Experience for Undergraduates program (NSF EEC 0851799, 1156549, and 1460776) at Kansas State University. We have had a team project each summer, which also included some other undergraduate students, related to the SPCSs research program. We thank all of these students and all others who helped with these team projects for their help to develop a better understanding of the issues related to advancing SPCSs.

Each summer CHE 670 Sustainability Seminar has been offered at Kansas State University. Many have helped with these seminars as speakers and in other ways to advance our understanding of the energy transitions that are taking place and the importance of SPCSs and EVs in the efforts to advance sustainable development and reduce greenhouse gas emissions. We thank all who have participated in these seminars and the annual Dialog on Sustainability.

Black and Veatch has provided funding for the project "Building a World of Difference with Solar Powered Charge Stations for Electric Vehicles" since 2012, and this funding has supported a number of students who have helped with research on SPCSs. We would like to thank Black and Veatch for this funding and thank Charles Pirkle, Kevin Miller, Forrest Terrell, and William Roush for their help.

We also acknowledge financial support through the Electric Power Affiliates Program and the leadership of Noel Schulz in this program and research related to electric power, smart grid, and decision support systems related to SPCSs, EVs, and other related topics.

The research program on SPCSs has had the benefit of input from a network of participants in the Consortium for Environmental Stewardship and Sustainability (CESAS). We thank all who have helped through CESAS.

In addition to those who are listed as authors in the book, we thank Darwin Abbott, Placidus Amama, Jennifer Anthony, Jack Carlson, Danita Deters,

Bill Dorsett, Keith Hohn, Jun Li, Ruth Miller, Behrooz Mirafzal, Bala Natarajan, John Schlup, Florence Sperman, and Sheree Walsh for their help.

Irma Britton has provided many ideas that have been valuable as we have attempted to prepare this manuscript for publication. We thank her for this.

The quotes that are included at the beginning of each chapter are taken from BrainyQuotes, Goodreads, and Phil Harding Quotes Corner. We thank them for having many good quotes to consider.

Larry E. Erickson
Jessica Robinson
Gary Brase
Jackson Cutsor

Contributors

Michael Babcock is professor of economics at Kansas State University. His research includes adoption rates of electric powered vehicles and he has received several national awards for research excellence in transportation economics.

Gary Brase is professor of psychological sciences at Kansas State University. His research includes personal decision making processes.

Jackson Cutsor is an undergraduate student in electrical engineering at the University of Nebraska-Lincoln who helped with the research and book while he was at Kansas State University in the summer of 2015.

Larry E. Erickson is professor of chemical engineering and director of the Center for Hazardous Substance Research at Kansas State University. He is one of the principal investigators on the NSF REU award and the Black and Veatch award (see Acknowledgments).

Ronaldo Maghirang is professor of biological and agricultural engineering at Kansas State University. His research is on air quality.

Anil Pahwa is professor of electrical and computer engineering at Kansas State University. His research includes electric power systems. He is a principal investigator on the Black and Veatch award and the Electric Power Affiliates Program award (see Acknowledgments).

Matthew Reynolds is an undergraduate student in chemical engineering at Kansas State University who helped with the research and book during the summer of 2014 and during the academic year since 2014.

Jessica Robinson is an undergraduate student at the University of North Carolina who helped with the research and book in the summers of 2014 and 2015 and the fall and winter of 2015.

Blake Ronnebaum is an undergraduate student in chemical engineering at Kansas State University who helped with the research and book in the summer of 2014 and in the fall of 2015.

Rachel Walker is an undergraduate student in chemical engineering at Kansas State University who helped with the research and book during the summer of 2015.

Andrey Znamensky is an undergraduate student in chemical engineering at Columbia University who helped with the research and book during the summer of 2015 while he was at Kansas State University.

1

Introduction

Larry E. Erickson, Gary Brase, Jackson Cutsor, and Jessica Robinson

CONTENTS

We cannot solve our problems with the same thinking we used when we created them.

Albert Einstein

There is an incredibly large and complex infrastructure built around transportation and fossil fuel power. This infrastructure includes thousands of oil fields, pipelines, huge refineries, and trucks to distribute gasoline to over 150,000 gasoline and service stations. There are over 250 million registered passenger vehicles in the United States and many more parking spaces. Personal vehicles in the United States consume more than 378 million gallons of gasoline every day, which is over 45% of the U.S. oil consumption according to the U.S. Energy Information Administration.

All that petroleum used for transportation is a major source of greenhouse gases, and on top of that are coal fired power plants that are a massive contributor of carbon dioxide emissions. In December 2014, at the United Nations COP 20 meeting in Lima, Peru, many delegates from nearly 200 nations signed an agreement to reduce greenhouse gas emissions. On December 12,

2015, the Paris Agreement on Climate Change was adopted by the Parties to the United Nations Framework Convention on Climate Change (UNFCCC, 2015). This agreement has a goal to reduce greenhouse gas emissions until carbon dioxide concentrations in the atmosphere stop increasing. The goal is to accomplish this balance of sinks and sources before 2100, but to begin as quickly as possible (UNFCCC, 2015). Similarly, the Clean Power Plan (U.S. EPA, 2015) calls for more electricity to come from renewable resources. The reduction of greenhouse gas emissions is one of the main goals of this plan. Doing that, though, means using less coal and petroleum. One of the great sustainability challenges is to increase the fraction of energy that comes from renewable resources. The finite supplies of fossil fuels and the greenhouse gas emissions associated with their combustion are important reasons to develop new technologies that allow progress in sustainable development. The goal of reducing greenhouse gas emissions by 80% by 2050 is considered to be appropriate, but how can we get there? To help accomplish this, it is important to electrify transportation and generate a significant fraction of electricity using renewable resources and nuclear energy (Williams et al., 2012). The transition to electric vehicles (EVs) and the construction of solar powered charging stations (SPCSs) to provide an infrastructure for EVs do go a long way toward accomplishing this. It can help generate more of our power needs from renewable resources and reduce greenhouse gas emissions and petroleum use.

Climate change is a "super wicked problem" because it is global, it affects everyone, and it involves entire ecosystems (Walsh, 2015). Climate change must be addressed because it has many impacts on our lives. Because action is needed in all countries, it is very difficult to find good solutions and implement them. The policy challenges associated with passing legislation and agreeing on regulations are "super wicked problems" because of potential impacts and global reach. The world needs research and development of new technologies that enable us to transition to a good life with an 80% reduction in greenhouse gas emissions and ample supplies of raw materials for future generations. Air quality will be improved as well.

1.1 Solar Power and Electric Vehicles

This book is about the sizeable challenges and the even greater opportunities offered by the marriage of solar power and electric vehicles (EVs) to provide an infrastructure for EVs. Strong and compelling cases can be made for adoption of EVs and a transition to sustainable energy.

An EV is much more efficient than a similar vehicle powered by gasoline. The EV is simple to construct because no engine cooling system is needed, no lubrication system is needed, there is no transmission, no exhaust system, and no catalytic converter is required. Maintenance costs are low. The space needed for the engine is small.

Strong and compelling cases can be made for sustainable energy, especially wind and solar energy. Solar power is growing rapidly. Lester Brown and colleagues (2015) have written about the great transition that has started from fossil fuels to wind and solar energy for electric power. The prices of wind and solar energy have decreased, and there is rapid growth in both technologies. Solar power production has been quietly getting more and more efficient, to the point where it is now as economically viable as other forms of producing electricity in many locations (Brown et al., 2015). We are already seeing rapid growth in distributed solar power generation in Europe and many other parts of the world.

Putting solar power and EVs together, we get an interaction effect that is beneficial to both; that is, the two technologies magnify the effects of each other because the batteries in EVs can store the clean energy produced by the solar panels. Because the batteries in EVs can store energy and EV owners can decide to charge when power costs are low, EVs can be beneficial to a power grid with wind and solar energy production and time-of-use prices for electricity.

1.2 Solar Powered Charging Stations (SPCSs)

One infrastructure alternative is to construct solar powered charging stations (SPCSs) in parking lots to produce electric power that flows into the electrical power grid. Covering 200 million parking spaces with solar panel canopies has the potential to generate 1/4 to 1/3 of the *total* electricity that was produced in 2014 in the United States. Even parking under the solar panel canopy has benefits, including shade and shelter from rain and snow. Meanwhile, the electrical grid can be used to charge the batteries of EVs.

Consider a world with a smart grid, millions of EVs, primarily powered by solar and wind energy, with millions of SPCSs and reduced emissions from combustion of coal and petroleum. What would it look like? Many countries can have energy independence with wind and solar power and EVs (a political goal for the United States since at least the Nixon administration). People would spend less on fuel (energy) and vehicle maintenance. The cleaner air would have social value and improve health.

The transformation to electric powered vehicles supported by an infrastructure of SPCSs and a smart grid will take some time because of the useful life of automobiles and electrical power generating plants. But recent progress in the development of solar panels and batteries has made this transformation possible. As the prices of solar panels and batteries for EVs decrease because of research and development, the rate of this transformation will increase. Many more individuals will purchase an EV as they realize that the cost of transportation is lower and more convenient with an EV than with a gasoline powered vehicle.

The number of new installations of solar panels to generate electricity has been growing rapidly. Between 2015 and 2050, progress in sustainable development may include the addition of many millions of EVs and SPCSs as well as installation of a smart grid with real time prices for electricity. The majority of vehicles sold in 2050 may be plug-in models; Toyota announced on October 14, 2015 that it aims to reduce the mass of carbon dioxide emitted from its new automobiles by 90% by 2050 (Japan for Sustainability, 2015). These anticipated developments have the potential to reduce greenhouse gas emissions substantially and create many jobs.

1.3 Air Quality

Air quality in urban areas will improve because EVs have no emissions when powered by electricity that is generated by solar energy. The improvement of urban air quality has social, environmental, economic, and health benefits. The quality of urban life would be much better in many cities of the world if all transportation was with EVs and these vehicles were powered with wind and solar energy.

The cost of gasoline will be lower because of the reduced demand as the number of EVs increases. Gasoline prices decreased in late 2014 because of increased supplies and the reduced demand. Part of that was the fact that more than 300,000 EVs were purchased and placed in service in 2014 worldwide, and this relationship can get stronger with more EV purchases.

1.4 Battery Storage and Infrastructure

The batteries in EVs are currently expensive, but they are important because they store the energy that is needed for travel in an EV. A large network of charge stations that allows EVs to be charged wherever they are parked would have significant value for EV owners. The size of the battery pack in an EV and the charging infrastructure are related because an EV owner can use that vehicle for many more purposes if a comprehensive supporting infrastructure is available and convenient. For example, an EV with a range of 85 miles (137 km) can be used for travel to and from work when the commuting distance is 50 miles each way if there is an infrastructure to charge the EV at work. An extensive charging infrastructure gives EV owners greater choice and convenience as to when and where to charge their EV.

This is important because electric power production and use need to be balanced when there is limited or no storage as part of the electrical grid. If the only place to charge an EV is at home, then there is a greater need to charge the battery when arriving home so it will be ready for the next trip. This may result, for example, in a significant number of EV batteries being charged after work at 5:30 p.m. on hot days when the load on the electrical grid is already near its maximum capacity.

A high availability of SPCSs aids in distributing demand on the electrical grid. Finally, as EV battery sizes increase EV range, it enables EVs to travel farther distances before requiring a charge, and it reduces the frequency in which EVs must visit charge stations.

1.5 Employment

The construction of the SPCSs and the modernization of the grid will provide construction and electrical jobs where the SPCSs are located and technical employment for those who install smart grid systems. There will also be employment associated with the equipment and materials that are used to construct the SPCSs and manufacture the smart grid equipment. Solar panels, inverters, smart meters, software, structural materials, communication equipment, and charge stations are needed.

1.6 Trillion Dollar Research Challenge

One of the important potential developments for EVs is less expensive batteries in terms of the cost per kWh of storage or cost per mile of range. Many current EVs have an efficiency of about 3 miles (5 km) per kWh. Battery costs in 2015 are about $300/kWh of capacity or $100/mile of range (Nykvist and Nilsson, 2015). A reduction in cost by 1/3 would have more than $1 trillion in value to society and make EVs less expensive by $500 to $10,000 depending on the size of the battery pack.

1.7 Real Time Prices for Electricity

There are many aspects associated with developing a solar powered charging infrastructure for EVs. The electrical power that flows into the grid should be properly valued and used. Real time prices or time-of-use rates are beneficial for EVs, SPCSs, and the electrical grid. Real time prices reflect the current demand on the electrical grid. Thus, peak power times have higher electricity prices. These pricing strategies can influence when vehicle owners charge their vehicles. Solar panels produce electricity during the day, when the value of power is higher than the average value. There are many opportunities to charge batteries in EVs when the demand for electricity is low and night time charging has been shown to be beneficial to utilities and EV owners in many locations with time-of-use prices. A large number of EVs with battery storage capacity changes the dynamics of the electrical energy network because substantial energy storage is available and prices can be used to encourage charging when surplus power needs to be stored. Grid modernization, though, is necessary to have effective communication and real time prices.

1.8 Shaded Parking

One of the significant aspects of adding SPCSs to parking lots is that shade is provided. It is more pleasant to enter a car that is in the shade on a hot sunny day, and the resale value of a car is better if it has been consistently parked in the shade. Adding solar panels above parking spaces requires very little additional land. Thus, SPCSs as a renewable energy alternative compares well with ethanol and wind energy in terms of land requirements.

1.9 Business Models for SPCS and EV Charging

Appropriate business models and permits are needed for SPCSs because electrical energy is regulated in many locations. Multiple parties (parking lot owner, charge station owner, utility, employer, vehicle owner) may be involved. How is the cost of the SPCS infrastructure to be paid for? Who makes a profit from EVs and SPCSs? What is the role of government policy? There are many social, environmental, economic, and policy aspects to consider. The convenience of charge stations is important for many people. Since the cost of electricity to drive 10 miles is of the order of $0.50 and the value of the electricity from charging with level 1 for two hours is less than $1.00, business models such as free parking that includes free charging are fairly common. The cost of the SPCS infrastructure can be paid for through sales income, taxes, or user fees. If there are no financial transactions associated with charging, it is convenient and efficient. These topics will be considered in later chapters.

1.10 Economic Externalities

The economics associated with the charging infrastructure of EVs include some positive externalities (benefits enjoyed by others, indirectly), because the costs of mitigating climate change and improving urban air quality can be included. This may help to spur some of the policy decisions that are needed to reach the goal of 80% reduction of greenhouse gas emissions by 2050. For instance, Saari et al. (2015) have investigated air quality co-benefits associated with a reduction of greenhouse gas emissions. When the benefits of climate change mitigation and improved air quality associated with the electrification of transportation are included, the value of an infrastructure of SPCSs is enhanced significantly.

1.11 Challenges and Opportunities

There are a number of actions and ongoing efforts that are beneficial to the goals of reducing greenhouse gas emissions and developing an infrastructure of SPCSs for EVs. These include:

1. Research to reduce cost and increase efficiency of solar panels.
2. Research to improve batteries and reduce their cost.
3. Progress in smart grid development and implementation including time-of-use prices.
4. Progress in developing approved procedures for electric utilities to install SPCSs and receive income as a regulated utility.
5. Public education on the benefits of the transformation to renewable energy, a smart grid, SPCSs, and EVs.

These actions are important and they will be discussed further in later chapters.

1.12 Sustainable Development

Sachs (2015) has pointed out that sustainable development is a science of complex systems. The complexity associated with the topics in this book arises because of the importance of environmental sustainability; the interactions of the world economy, global society, and the environment; and the difficulty in making optimal decisions where utilities are regulated and there are important economic externalities. A modernized smart electrical grid with large amounts of wind and solar energy adds complexity because of variations in solar radiation and wind speed. Battery storage has the potential to be very helpful in grid design and operation, but there are complexity issues associated with a smart grid that includes these renewable sources and battery storage in EVs that are controlled by customers who may respond to real time prices.

1.13 Objectives of the Book

One objective of this book is to describe pathways and challenges to go from our present situation to a world with a better, sustainable transportation system: one with EVs, SPCSs, a smart grid with real time prices, more energy storage, reduced greenhouse gas emissions, better urban air quality, abundant wind and solar energy, and electricity for all who live on this planet. Because the topics of the chapters are complex, there is some consideration of related topics across various chapters.

At a broad level, in order to have good governance in the world we need to have educated people making good decisions. This book introduces important topics and provides information that will be helpful to decision makers, engineers, public officials, entrepreneurs, faculty, students, and members of organizations that work cooperatively to make this a better world.

At a more personal level, another objective of this book is to provide encouragement and knowledge that will be helpful to those who wish to own an EV and an SPCS. Many readers will be involved in smart grid modernization accompanied by variable prices, and some understanding of the benefits associated with time-of-use and real time prices will be helpful to them.

References

Brown, L.R., J. Larson, J.M. Roney, and E.A. Adams. 2015. *The Great Transition: Shifting from Fossil Fuels to Solar and Wind Energy*, W.W. Norton & Co., New York.

Japan for Sustainability. 2015. Toyota announces 'Environmental Challenge 2050,' *Japan for Sustainability Weekly*, December 1–7, 2015, http://www.japanfs.org/.

Nykvist, B. and M. Nilsson. 2015. Rapidly falling costs for battery packs for electric vehicles, *Nature Climate Change*, 5: 329–332.

Saari, R.K., N.E. Selin, S. Rausch, and T.M. Thompson. 2015. A self consistent method to assess air quality co-benefits from U.S. climate policies, *Journal of Air and Waste Management Association*, 65: 74–89.

Sachs, J. 2015. *The Age of Sustainable Development*, Columbia University Press, New York.

UNFCCC. 2015. Paris Agreement, United Nations Framework Convention on Climate Change, FCCC/CP/2015/L.9, December 12, 2015, http://unfccc.int/.

U.S. EPA. 2015. Carbon pollution emission guidelines for existing stationary sources: Electric utility generating units, U.S. EPA: http://www.epa.gov/.

Walsh, B. 2015. President Barack Obama takes the lead on climate change, *Time*, August 17, 2015.

Williams, J.H., A. DeBenedictis, R. Ghanadan, A. Mahone, J. Moore, W.R. Morrow III, S. Price, and M.S. Torn. 2012. The technology path to deep greenhouse gas emission cuts by 2050: The pivital role of electricity, *Science*, 335: 53–59.

2

Electric Vehicles

Rachel Walker, Larry E. Erickson, and Jackson Cutsor

CONTENTS

If I had asked people what they wanted, they would have said faster horses.

Henry Ford

2.1 Introduction

An electric vehicle (EV) has the advantage of being very simple to design and build. The EV is very efficient particularly in comparison to internal combustion engine vehicles (ICEs); there is no radiator and engine cooling system that uses fluids in most EVs. Since there are no exhaust emissions, no catalytic converter is needed. This simplicity reduces maintenance costs. In the last several years, many new EVs have been introduced and made available for sale in the United States and throughout the world (Inside EVs, 2016). More than 500,000 EVs were manufactured and delivered in 2015 in the world (Inside EVs, 2016).

One example of an all-electric vehicle is the Tesla S. Powered by either a dual or single electric motor depending on the model, the Tesla S has a range of 240–270 miles at full charge. It runs on a 70–85 kilowatt hour (kWh) battery, comes with an eight-year battery and drive unit warranty, and gives purchasers a $7500 federal tax credit. Tesla provides free charging to Tesla owners via its Supercharge network of charging stations located throughout the country. This vehicle saves owners an estimated $10,000 in gas over a five-year period (Tesla Motors, 2015).

Extended range electric vehicles (EREVs) are powered by an electric motor, but also contain a gasoline engine that powers a generator that charges the batteries in the vehicle. The Chevrolet Volt is an example of an EREV. The 2015 Volt has an estimated gas-free range of 38 miles when fully charged. With a fully charged battery and a full tank of gas, the 2015 Volt's range becomes approximately 380 miles. The 2016 Volt has a range of 50 electric miles from its batteries (Chevrolet, 2015a).

A third type of electric vehicle is the plug-in hybrid electric vehicle (PHEV), such as the Toyota Plug-in Prius. This type of vehicle can operate as an electric vehicle as long as there is sufficient energy in the battery, and it can operate using both gasoline and electricity. When the battery is low, the PHEV performs the same as a Prius hybrid that does not have a plug-in connection. It makes use of both the electrical drive system and an internal combustion engine with the gasoline motor turning off when stopped at stoplights. Plug-in Prius buyers receive an estimated tax credit of $2500 (U.S. Department of Energy, 2015b).

Owning an EV can be very advantageous for drivers. The simple design, low maintenance costs, efficiency, convenience of home charging, and environmental benefits make EVs a competitive option. Disadvantages include short driving ranges, higher purchase price, heavier vehicle weight, large batteries, and inconvenience and expense of charging vehicles when away from home. While researchers work to find solutions to these drawbacks, plans to increase EV sales and push the United States in an environmentally beneficial direction continue.

This chapter will include details on the first EVs invented, current developments in EV research, and the design of each type of EV. It will also give information about particular EV models, efficient features specific to EVs, and current sales throughout the United States and the world. This chapter will show readers many environmental and financial incentives for EV buyers, including government policy incentives, and will explore life cycle analysis of EVs versus combustion engine vehicles.

2.2 History of EVs

EVs have been in existence since the nineteenth century, but have not been a realistic option for everyday travel until recently. Europeans were the first to experiment with making EVs, but the United States was close to follow. In 1890, William Morrison, a chemist from Des Moines, Iowa, created the first EV in the United States (Matulka, 2014). By 1900, EVs were very popular

(Matulka, 2014). At this time, steam and gasoline powered vehicles were limited in range and took manual time and effort to start (Matulka, 2014). EVs were quieter and easier to drive, which made them ideal for short drives within cities (Matulka, 2014). However, developments in ICEs and increased availability of gasoline put an end to the brief prominence of EVs. As these advances in technology continued, the EV was no longer a competitive option (Matulka, 2014).

Until the early 1990s, no real progress or attempts were made at revitalizing the concept of an EV. In 1996, General Motors released the EV1, a small car that was completely electric; see Figure 2.1. Even though there was almost no charging infrastructure and the range was a maximum of 100 miles, it was met with considerable enthusiasm from the public, especially in California. Although there was clear public support, GM received much negative pressure from corporations and developed concerns that the EV1 would have a negative effect on the automobile industry. Despite owner protest, GM decided to remove them from the market. They recalled and crushed all of just over 1000 EV1s (General Motors EV1, 2015a). However, General Motors has shown renewed support for EVs with its recent announcement in 2015 that it will be producing a new all-electric vehicle with a proposed range of more than 200 miles (Chevrolet, 2015b).

A number of different factors have led to the recent increase in EV and PHEV production, including government support, environmental concerns, new technology, and the projected increase in the price of operating an ICE. The corporate average fuel economy (CAFE) regulations provide an incentive for manufacturers to market EVs and PHEVs. Government subsidies at the federal and state level have made EVs more attractive by giving owners a significant tax break.

Recent years have shown the need for a more sustainable transportation option. Not only do ICEs drive a U.S. dependence on foreign oil, but they

FIGURE 2.1
Pictured is the 1996 General Motors EV1. (Photo from Henry Ford Blog. General Motors' EV1. *The Henry Ford Blog*. n.p., June 22, 2015. Web. Jan. 14, 2016. http://blog.thehenryford.org/.)

also release exhaust pollutants and evaporative emissions that are harmful to the environment (Environmental Protection Agency, 2012). Despite efforts to reduce these emissions, such as the Clean Air Act of 1970, the problem continues to grow because the number of miles people drive has more than doubled since this act was passed (Environmental Protection Agency, 2012). As a result, the government has approved initiatives to increase research to make EVs a more efficient, financially viable option.

2.3 Features of EVs

The EV is powered by at least one electric motor, which is fueled by rechargeable battery packs (U.S. Department of Energy, 2015a). It produces no greenhouse gas emissions and generally the batteries can be recharged in a matter of hours (Berman, 2014). Because they do not have an internal combustion engine, EVs do not need the level of maintenance that ICEs require (Berman, 2014). EVs also operate much more quietly.

EVs operate with a higher efficiency level than gasoline-powered vehicles. In fact, 59–62% of electrical grid energy is converted to power at the wheels by EVs as opposed to 17–21% converted by gasoline-powered vehicles (U.S. Department of Energy, 2015a). EVs also have the potential to reduce energy dependence, since electric energy can be generated domestically (U.S. Department of Energy, 2015a).

Many EVs are also equipped with regenerative braking, a system allowing the kinetic energy associated with braking to be stored in the car batteries or super capacitors. This energy can then be used to extend the range of the EV (Lampton, 2009). Some examples of EVs equipped with regenerative braking capabilities are the Nissan Leaf, Toyota Prius, Chevrolet Volt, and Tesla Roadster.

2.4 Charging EVs

When it comes to charging an EV, there are several options available. First, there are two common types of charging: Levels 1 and 2. A Level 1 charger connects to a 120-volt power source; this is the energy level of most outlets found in homes throughout the United States. According to the U.S. Department of Energy, "Level 1 charging, which adds about 6 miles of electric-drive range per hour of charging, may be a suitable option for those with shorter commutes or for those who can leave their vehicle plugged in for an extended period of time" (Lutterman, 2013). Level 2 charging takes place through a 240-volt outlet, which requires EV owners to buy and install the necessary equipment if they wish to have Level 2 charging at home. Level 2 charging is much faster than Level 1; it can add approximately 10 to 20 miles of EV drive range per hour charged (Lutterman, 2013).

Many other EV charging options exist outside the home. As of July 9, 2015, there were 9974 charging stations and 25,934 electric outlets publicly available in the United States. California leads as the state with the most charging stations (2214) and electric outlets (7375) (U.S. Department of Energy, 2015c). Tesla provides a network of Supercharge stations that are available throughout the country for Tesla EVs. These Superchargers, which charge even faster than Level 2 chargers, are cost-free but only available to Tesla drivers.

As of 2015, Volta Industries has partnered with companies to offer EV charging available for all EVs, paid for by advertising shown at charging stations while vehicles charge (Volta Charging, 2015; Wang, 2015). These are a few examples of public charging available to EV owners. Charging infrastructure continues to expand worldwide as EV adoption grows.

2.5 Current EVs on the Market

There are many EVs currently for sale in the United States and worldwide. Generally, these vehicles are small and offer limited seating due to large, heavy battery packs. They also have limited all-electric driving ranges. Researchers continue to find ways to design EVs that can compete with every type of ICE. Table 2.1 lists many popular EVs on the market in July 2015.

TABLE 2.1

Reported Prices, All-Electric Range, and Battery Size of Some Plug-In Vehicles, July 2015[a]

Vehicle	Price (US Dollars)	Battery Size (kWh)	All-Electric Range (Miles)	Type of Vehicle
Chevrolet Volt	$34,170	17.1	38	EREV
Ford C-Max Energi	$31,770	7.6	21	PHEV
Ford Focus	$29,170	23	76	EV
Ford Fusion Energi	$35,525	7.6	20	PHEV
Honda Accord PHEV	$39,780	6.7	13	PHEV
Mercedes-Benz B-Class Electric	$41,450	28	84	EV
Nissan Leaf	$29,010	24	84	EV
Tesla S	$75,000–105,000	70–85	240–270	EV
Toyota Plug-In Prius	$31,184	4.4	11	PHEV

[a] Information from company Internet sites.

As of May 2015, the best-selling EV in the United States was the Nissan Leaf, followed by the Tesla S, and then the Chevy Volt (Shahan, 2015a). In 2015, U.S. sales reached 43,973 by May (Cole, 2015); they were more than 116,000 for the year (Inside EVs, 2016).

Various passenger EVs are currently being developed, including EV pick-up trucks and mini-vans. One example of a multi-passenger EV is the 2016 Volvo XC90 T8, a luxury hybrid plug-in SUV with an expected range of at least 96 miles (Voelcker, 2015). Additional forms of EV transportation are developing in the forms of electric bikes and scooters as well. Even with these developments, a wider variety of EV choices is needed to meet customer needs. For example, few affordable family size EVs are currently available on the U.S. market. Audi intends to market a family size SUV, starting in 2018 (Collie, 2015). Mitsubishi is selling the family size Mitsubishi Outlander PHEV in Japan and Europe and plans to market this vehicle in the United States in 2016 (Mitsubishi, 2015; Shahan, 2015b). It is selling in large numbers in Europe (Shahan, 2015b).

Chevrolet has introduced the 2017 Chevrolet Bolt, which is an EV with a range of more than 200 miles and a projected price of less than $30,000 after government tax credits have been deducted (Bell, 2016). Tesla Motors is also planning to manufacture an EV with a range of more than 200 miles that will be in the same price range as the Bolt. Tesla Motors sold over 50,000 EVs in 2015, and hopes to sell about 500,000 EVs in 2020 (Zhang, 2015; Waters, 2016).

In the future, marketing efforts should be made to increase EV sales. Customers should be educated on the environmental benefits and overall efficiency of EVs. When discouraged by high retail prices, car buyers should look at the long-term cost of an ICE including maintenance and fuel versus an EV (Telleen and Trigg, 2013). EV charging infrastructure, a vital component of making EVs practical and competitive with ICEs, is continuing to develop as new charge stations are designed and built throughout the United States. These new developments will allow EVs to become more marketable for a wider range of customers.

More than one million EVs are now in use in the world (Shahan, 2015c). EV sales were more than 39% higher in 2015 compared to 2014 in the world (Inside EVs, 2016).

2.6 Environmental and Economic Benefits

At present, there are many incentives for customers to buy an EV. On a grand scale, policy, environmental, economic, and social issues drive EV research and development within the United States and worldwide. These include government energy standards, tax incentives, environmental benefits, and political initiatives. From an ecological standpoint, these issues include incentives with social value, a critical concern for the future of the environment, and the reduction of greenhouse gas emissions.

In the United States, the national government pushes efforts to reduce emissions in many ways. The CAFE standards, created by Congress in 1975, continually set new gas mileage and fuel standards to "reduce energy consumption by increasing the fuel economy of cars and light trucks" (National Highway Traffic Safety Administration, 2015). According to the National Highway Traffic Safety Administration, "The proposed standards are expected to lower CO_2 emissions by approximately 1 billion metric tons, cut fuel costs by about $170 billion, and reduce oil consumption by up to 1.8 billion barrels over the lifetime of the vehicles sold under the program. These reductions are nearly equal to the greenhouse gas (GHG) emissions associated with energy use by all U.S. residences in one year" (National Highway Traffic Safety Administration, 2015).

The U.S. federal government currently (July 2015) offers tax credits to those who purchase EVs. For example, a $7500 federal tax credit is currently offered for purchase of 22 different EV models, including the Nissan Leaf and the Tesla Model S (U.S. Department of Energy, 2015b). Federal tax credits are also available to purchasers of 16 different PHEV models, ranging from a $2500 credit with the Toyota Prius Plug-in Hybrid to $7500 for the Chevrolet Volt (U.S. Department of Energy, 2015b). Additional tax credits vary from state to state.

Through life cycle analysis (LCA), total energy input and output can be measured for ICEs, EVs, and PHEVs. According to an LCA done through the University of California, Los Angeles, the lifetime energy requirements of ICEs are far higher than those of EVs and PHEVs. Specifically, over its lifetime, an ICE requires 858,145 MJ (mega-joules) of energy; the EV, 506,988 MJ; and the PHEV, 564,251 MJ. This LCA also compares lifetime CO_2 emissions of each vehicle. Data shows the ICE releases 0.35 kg CO_2eq/mile; the EV, 0.18; and the PHEV, 0.23. It is important to keep in mind that "the use phase can be attributed to 96% of ICE emissions, 91% of PHEV emissions, and 69% of EV emissions. Battery manufacturing is accountable for 24% of the [EV's] lifecycle emissions, but only 3% of hybrid's lifecycle emissions" (Aguirre et al., 2012). For plug-in vehicles, the CO_2 emissions depend on how the electricity was generated.

The federal government has taken specific steps to promote EV use in order to combat environmental harm. In 2012, President Barack Obama released an initiative through the U.S. Department of Energy called the EV Everywhere Grand Challenge. This initiative focuses on U.S. advancement of EV technology to make EVs as affordable for the average American family by 2022 as a 2012 baseline gasoline-powered vehicle. Its blueprint specifically outlines vehicle weight reduction by nearly 30%, electric drive system cost reduction from $30/kW to $8/kW, and battery cost reduction from $500/kWh to $125/kWh (U.S. Department of Energy, 2013). EV Everywhere focuses on technological developments as well as federal and state support and policy to achieve its goal (U.S. Department of Energy, 2013). As of January 2014, battery costs had been reduced to $325/kWh and a $5/kW electric drive system had been developed (U.S. Department of Energy, 2014). Through continued research and outreach, EV Everywhere continues to progress rapidly toward and beyond its target advancements (U.S. Department of Energy, 2014).

On an international level, several governments from countries around the world have worked together to form the Electric Vehicles Initiative (EVI) (Telleen and Trigg, 2013). This initiative was launched in 2010 under the Clean Energy Ministerial, a dialogue between countries. EVI encourages a worldwide EV adoption goal by 2020 and specifically outlines the action necessary, such as government action, infrastructure, technology, and marketing (Telleen and Trigg, 2013).

In addition to policy and environmental incentives, world leaders have brought recent attention to the importance of sustainable energy. This educates the public and provides more incentive for drivers to choose to buy EVs. In particular, Pope Francis brought ecological issues to attention publicly through his encyclical letter on climate change:

> Humanity is called to recognize the need for changes of lifestyle, production and consumption, in order to combat this warming or at least the human causes which produce or aggravate it. It is true that there are other factors (such as volcanic activity, variations in the earth's orbit and axis, the solar cycle), yet a number of scientific studies indicate that most global warming in recent decades is due to the great concentration of greenhouse gases (carbon dioxide, methane, nitrogen oxides, and others) released mainly as a result of human activity. Concentrated in the atmosphere, these gases do not allow the warmth of the sun's rays reflected by the earth to be dispersed in space. The problem is aggravated by a model of development based on the intensive use of fossil fuels, which is at the heart of the worldwide energy system. (Francis, 2015)

Pope Francis addressed climate change as a moral issue. He specifically pointed out the urgent need for people throughout the world to address air pollution and consumption of nonrenewable resources; these are issues that are directly addressed by EV research (Francis, 2015). This encyclical has reached the political world; California Governor Jerry Brown, a pollution prevention advocate, acknowledged the papal responsiveness to environmental issues. Brown stated, "It's now up to leaders in business and government—and wherever else—to join together and reverse our accelerating slide into climate disorder and widespread suffering" (Jennewein, 2015).

2.7 EV Disadvantages and Challenges

Many challenges stand in the way of making EVs a competitive option for all drivers. EVs are less efficient in the winter when energy is used to heat the cabin and defrost the windshield. For ICEs, waste engine heat is used for these purposes. Since temperature affects battery performance, the range of EVs is reduced when the temperature is low in cold environments. The ambient temperature where the EV is parked affects the amount of charge that the battery is able to store. The drawbacks to EVs include a more limited driving range and longer charging time versus fueling time. Battery packs are also expensive to replace. Although EVs themselves produce no tailpipe emissions, power plants that provide the cars with electric energy may produce pollutants. Researchers are addressing these issues by finding new ways to increase battery storage and decrease charging time and costs. Further challenges include high retail prices, lack of policy and political initiatives, consumer education, and marketing.

One of the biggest roadblocks to EV adoption is the technology. As outlined in the Electric Vehicle Initiative (EVI),

> the most significant technological challenges currently facing electric-drive vehicles are the cost and performance of their components, particularly the battery. Price per usable kilowatt hour of a lithium-ion battery ranges between $300–400 and thus makes up a large portion of a vehicle's cost, depending on the size of the battery pack (Nykvist and Nilsson, 2015). A Nissan LEAF, for example, has a 24 kWh battery that costs approximately $7200, which represents about a fourth of the vehicle's retail price. Similarly, Ford uses a battery that costs between $7200 and $9000 for its Focus Electric, an electric version of its gas-powered Focus that itself sells for around $22,000. (Telleen and Trigg, 2013)

Due to the range limitations, high retail costs, and inconsistent charging infrastructure, EVs are not yet as affordable and practical as ICEs in many contexts. However, battery costs have been decreasing with time, and EVs have a promising future. Their simple design, low greenhouse gas emissions, energy efficiency, and overall sustainability are attractive to consumers on a global level. As petroleum becomes more expensive and scarce, sustainable options like EV transportation will have to be considered. Researchers continue to work to find solutions to make batteries less expensive and more efficient; city planners work to design practical charging infrastructure; and the government continues to push for EV adoption through policy and financial incentives. Through these combined efforts, EVs can become a competitive transportation option.

References

Aguirre, K., L. Eisenhardt, C. Lim et al. 2012. Lifecycle analysis comparison of a battery electric vehicle and a conventional gasoline vehicle. California Air Resource Board. http://www.environment.ucla.edu/media/files/BatteryElectricVehicleLCA2012-rh-ptd.pdf (Accessed July 8, 2015).

Bell, K. 2016. 2017 Chevrolet Bolt EV: Production electric car unveiled at consumer electronics show, *Green Car Reports*; http://www.greencarreports.com/.
Berman, B. 2014. What is an electric car? Electric vehicles, plugin hybrids, EVs, PHEVs. http://www.plugincars.com/electric-cars (Accessed June 8, 2015).
Chevrolet. 2015a. http://www.chevrolet.com/volt-electric-car.html (Accessed July 25, 2015).
Chevrolet. 2015b. Chevrolet commits to Bolt EV production; http://www.chevrolet.com/.
Cole, J. 2015. May 2015 plug-in electric vehicle sales report card. http://insideevs.com/may-2015-plug-electric-vehicle-sales-report-card/ (Accessed July 25, 2015).
Collie, S. 2015. Audi electric SUV concept quick off the mark, over 300 mile range. *Gizmag*, September 15, 2015; http://www.gizmag.com.
Environmental Protection Agency. 2012. Automobile emissions: An overview. http://www.epa.gov/otaq/consumer/05-autos.pdf (Accessed June 3, 2015).
Francis I. 2015. Encyclical Letter Laudato si' http://w2.vatican.va/content/francesco/en/encyclicals/documents/papa-francesco_20150524_enciclica-laudato-si.html (Accessed July 8, 2015).
General Motors EV1. 2015a. General Motors EV1, Wikipedia; https://en.wikipedia.org/.
General Motors EV1. 2015b. The Henry Ford Blog. http://blog.thehenryford.org/ (Accessed January 14, 2016).
Inside EVs. 2016. Monthly plug-in sales scorecard, January 2016; http://insideevs.com/.
Jennewein, C. 2015. Brown hails Pope's controversial encyclical on climate change. http://timesofsandiego.com/tech/2015/06/20/brown-hails-popes-controversial-encyclical-on-climate-change/ (Accessed July 8, 2015).
Lampton, C. 2009. How regenerative braking works. http://auto.howstuffworks.com/auto-parts/brakes/brake-types/regenerative-braking.htm (Accessed August 6, 2015).
Lutterman, J. 2013. Charging your plug-in electric vehicle at home. http://energy.gov/energysaver/articles/charging-your-plug-electric-vehicle-home (Accessed July 8, 2015).
Matulka, R. 2014. U.S. Department of Energy. The history of the electric car. http://energy.gov/articles/history-electric-car (Accessed June 3, 2015).
Mitsubishi. 2015. Mistubishi Outlander PHEV; http://www.mitsubishicars.com/.
National Highway Traffic Safety Administration. 2015. CAFE—Fuel economy. http://www.nhtsa.gov/fuel-economy (Accessed June 25, 2015).
Nykvist, B. and M. Nilsson. 2015. Rapidly falling costs of battery packs for electric vehicles. *Nature Climate Change* 5: 329–332.
Shahan, Z. 2015a. US electric car sales—Top 3 on top again. http://evobsession.com/us-electric-car-sales-top-3-on-top-again/ (Accessed July 8, 2015).
Shahan, Z. 2015b. The most popular electric cars in Europe may surprise you, *Gas2*; July 16, 2015; http://gas2.org/.
Shahan, Z. 2015c. One million electric cars will be on the road in September, *Clean Technica*, August 8, 2015; http://cleantechnica.com.
Telleen, P. and T. Trigg. 2013. Global EV outlook: Understanding the electric vehicle landscape to 2020. https://www.iea.org/publications/globalevoutlook_2013.pdf (Accessed July 16, 2015).
Tesla Motors. 2015. http://www.teslamotors.com/models (Accessed July 25, 2015).
U.S. Department of Energy. 2013. EV Everywhere blueprint. http://energy.gov/sites/prod/files/2014/02/f8/eveverywhere_blueprint.pdf (Accessed June 25, 2015).

U.S. Department of Energy. 2014. EV Everywhere: Road to success. http://energy.gov/sites/prod/files/2014/02/f8/eveverywhere_road_to_success.pdf (Accessed June 25, 2015).
U.S. Department of Energy. 2015a. All-electric vehicles. https://www.fueleconomy.gov/feg/evtech.shtml (Accessed June 8, 2015).
U.S. Department of Energy. 2015b. Federal tax credit for electric vehicles purchased in or after 2010. https://www.fueleconomy.gov/feg/taxevb.shtml (Accessed June 25, 2015).
U.S. Department of Energy. 2015c. http://www.afdc.energy.gov/fuels/stations_counts.html (Accessed July 16, 2015).
Voelcker, J. 2015. 2016 Volvo XC90 T8 plug-in hybrid 'twin-engine': First drive. http://www.greencarreports.com/news/1096866_2016-volvo-xc90-t8-plug-in-hybrid-twin-engine-first-drive (Accessed August 11, 2015).
Volta Charging. 2015. http://voltacharging.com/home (Accessed July 8, 2015).
Wang, U. 2015. An EV charging startup raises $7.5M to give away electricity for free. http://www.forbes.com/sites/uciliawang/2015/06/10/5754/ (Accessed July 8, 2015).
Waters, R. 2016. Tesla sales pace falls short at end of 2015, Financial Times, January 3, 2016; http://www.ft.com.
Zhang, B. 2015. Elan Musk believes the Model X will double Tesla's sales, *Business Insider*, July 8, 2015; http://www.businessinsider.com.

3

Solar Powered Charging Stations

Larry E. Erickson, Jackson Cutsor, and Jessica Robinson

CONTENTS

When something is important enough, you do it even if the odds are not in your favor.

Elon Musk

Solar powered charging stations (SPCSs) are one of the important developments related to the electrification of transportation. The number of sites with SPCSs is increasing because of their value and convenience. In many cases, the SPCSs are designed to allow the electricity that is generated to flow into the local electrical grid. The solar panels provide shade in the parking lot, and the charge station is connected to the grid such that power for charging EVs is available at all times. At some sites there are batteries for electrical storage also. Some sites have battery storage without any grid connection. In cases where the power is provided to the EV without any cost to the owner of the EV, the charging equipment is simpler than when customers need to pay for connecting to the electric vehicle supply equipment (EVSE).

Many SPCSs have a concrete base, steel frames and supports, and needed electrical components including transformers, wires, and inverters. In many cases, there is a payment station with payment software and hardware and communication capabilities.

In some locations, there are solar panels in parking lots, but there are no charging stations for EVs. These structures have been put in place to produce electricity and provide shade. Some were put in place before there was a demand for EVSEs. In these cases, a decision was made to construct the system without considering the need for EVSE infrastructure for EVs. There are many locations where SPCSs can be used to increase the amount of power generated with sustainable energy at competitive prices. Adding sustainable energy to the electrical grid with SPCSs has value for society because it is a very clean source of energy. These sites can be easily equipped with EVSEs when there is a need for them.

Envision Solar International, Inc. (2015) has developed a solar powered charge station with battery storage that is designed to be self contained and not connected to the electrical grid. This electric vehicle autonomous renewable charger can be towed to the site and used immediately. It also can be moved to a new site easily. It has 22 kWh of battery storage, which allows about one day of energy storage. The 2.3 kW solar array generates approximately 16 kWh/day, and it has a solar tracker to allow the solar array to follow the sun. This system can be installed at locations where there is no grid such as in parks, trail heads, and along roads where tourists may wish to stop. See Figure 3.1.

FIGURE 3.1
Solar powered charging system with battery storage available from Envision Solar International. (Photo provided by Envision Solar International, Inc.)

The amount of power that flows from the solar panels over a parking space depends on location, area of the panels, and efficiency. For instance, in Kansas a reasonable estimate is 16 kWh/day for one parking space. If 200 million parking spaces are covered with solar panels, 3.2 billion kWh/day could be generated, which can be compared to 11.2 billion kWh generated in the entire United States on an average day (Erickson et al., 2015). There are more than 200 million vehicles in use in the United States, and there are many more parking spaces than vehicles because there are always many empty parking spaces at any given time. Sports stadiums, church parking lots, shopping centers, and many work sites have empty spaces in their parking lots at many times during the week. The available land for SPCSs, the potential reduction in greenhouse gas emissions, and the reduced use of water compared to alternatives are metrics that favor SPCSs.

This chapter provides an introduction to SPCSs, and it builds on earlier papers by Goldin et al. (2014) and Robinson et al. (2014). The SPCS is an ideal example of sustainable development and the application of the triple bottom line principle: There are social, environmental, and economic benefits associated with SPCSs.

3.1 Social Benefits of SPCSs

Social benefits include shade, better air quality, and convenience. There are personal comfort benefits associated with entering a vehicle that has been in the shade on a hot summer day. Goldin et al. (2014) point out that the temperature in a car that is in the shade on a hot day may be more than 50°F lower. The social value of better air quality because of EVs and SPCSs is a benefit that impacts everyone. Economically SPCSs provide construction and maintenance jobs and reduce travel costs.

The reduction of greenhouse gas emissions has global benefits while the improved urban air quality associated with the transition to EVs and SPCSs benefits everyone in the urban area. Quality of life issues are important to many people. For example, some people move to the edge of an urban area in order to have better air quality.

Convenience is of significant social value to many people. If EV owners are able to plug in when they arrive at their parking space at work, when they stop at the mall after work, and when they are at home, this will have value for them, especially if there is a need to charge the batteries at sites other than at home. Constructing SPCSs at many locations will improve convenience for many EV owners. This convenience may help to retain employees, attract customers to a store, health club, or restaurant, and encourage purchases of EVs.

3.2 Environmental Benefits of SPCSs

Environmental benefits include reduced greenhouse gas emissions, better air quality in urban environments, and less noise. The transition to SPCSs has global environmental benefits because of reduced greenhouse gas emissions. The global goal of reducing emissions by 80% by 2050 will require significant changes, including the electrification of transportation and the generation of most of the electricity using sustainable methods such as solar panels. The electricity generated by SPCSs does not have air emissions associated with it. Air quality is impacted by emissions associated with coal fire power plants. Combustion gases can be controlled; however, there are costs associated with this and pollutants that are removed from the air exhaust become pollutants in waste water in some cases. There are no significant water requirements associated with solar energy compared to electricity generated with coal, nuclear, and natural gas where cooling water is used and lost to the atmosphere. Petroleum, coal, and natural gas production have significant environmental impacts, risks of production level spills and contamination, water use is significant, pipelines for transportation may rupture, and coal trains may leave the tracks.

A phenomenon affecting large cities is the urban heat island effect. This occurs because of a lack of vegetation, massive quantities of heat-absorbing materials such as concrete, and tall buildings that alter wind patterns. All of these issues make cities one or more degrees centigrade warmer than the surrounding rural areas on average. The solar panels on buildings and on SPCSs take solar energy and convert it to electrical energy, much like plants take light energy and convert it to chemical energy. Since EVs are much more efficient compared to cars with internal combustion engines (ICEs), the amount of heat generated per mile traveled by transportation is reduced. Per mile traveled, the ICE uses about 3 to 4 times as much energy as an EV. These two factors reduce the heat island effect.

In the STAR Community Rating System (STAR, 2015), SPCSs and EVs help communities meet 12 of 44 objectives, including green infrastructure, ambient noise, green market development, greenhouse gas mitigation, resource efficient public infrastructure, and greening the energy supply. STAR refers to Sustainability Tools for Assessing and Rating communities, and the STAR system is helpful to communities that want to track their progress toward a number of sustainability objectives.

3.3 Economic Benefits

Economically, SPCSs are beneficial on both a local and national level. They create temporary construction jobs and employment for those who produce the materials and parts that are used for the construction of the SPCS. There is also employment for those who manage and maintain the SPCSs.

Businesses, especially those with large fleets of vehicles, have the potential to save money by investing in SPCSs and EVs. Delivery vehicles can be drastically cheaper to operate with electrical power and with SPCSs can potentially be free to fuel after the initial investment has been paid off. The operational cost is about 33%–50% of a conventional vehicle if maintenance costs are included. The U.S. Postal Service could save on operational costs by using EVs and SPCSs. Since the EV does not use much power while it is stopped, it is especially efficient for mail delivery. Businesses have other reasons to invest, like the green halo effect and employee retention. Free charging while at work is an inexpensive benefit for a company to provide. People respect businesses that are ecofriendly, and this may help attract and keep customers, especially those who appreciate free charging while at the business.

The operating and maintenance costs of an EV are less than for an auto with an internal combustion engine. Goldin et al. (2014) reported that the cost of transportation is least for the Nissan Leaf EV when it is compared to several other vehicles. If SPCSs allow an individual to use a Leaf to come to work, this has economic value because transportation costs are reduced. When it is powered by electricity from solar energy, the Leaf is a very clean form of transportation, and this has economic value because the improved air quality reduces health costs in urban areas where air quality is impacted by transportation emissions. The economic benefits include the greater value a vehicle has as a used vehicle when it has been sheltered from the sun regularly. Battery life in EVs may be impacted by high temperature, and shaded parking may be beneficial on hot summer days. In the future, solar panel costs and battery costs are expected to be less than they are today. Simple, inexpensive electric vehicles will have great utility in many parts of the world, especially if they can be supported by SPCSs at many locations. For instance, Jordan is one of the countries that are moving forward with EVs and SPCSs (Ajumni, 2015).

3.4 Electric Vehicle Supply Equipment

The equipment that is used to charge electric vehicles includes Level 1, Level 2, and high rate EVSE (USDOE, 2013). Level 1 EVSE is for use with a 120 volt AC circuit. Most EVs are supplied with a Level 1 charging cord that has an automatic stop to terminate charging when the battery is charged. There is a standard 120 volt three-prong household plug on one end and a standard connector that plugs into the vehicle on the other end. Level 1 charging often adds about 5 miles of range or about 2 kWh per hour to the batteries. This rate of charging is about equal to the rate of supply of the solar panels above one parking space.

Level 2 EVSE uses a 240 volt supply often with a dedicated 40 amp circuit to provide approximately 18 miles of range or about 6 kWh per hour to the batteries. In many cases, the connection to the power supply is hard wired for safety. It is connected to the vehicle with the same J1772 standard connector as is used for Level 1 charging. The rate of charging depends on the charger that is in the vehicle. A 30 amp rate is commonly used.

Level 3 EVSE is often identified as DC fast charging and it is not as standardized as Level 1 and Level 2. Some EVs such as the Nissan Leaf that are equipped to accept DC fast charging have the CHAdeMO connector (Herron, 2015). There is also the SAE Combo Charging System (SAE CCS), which is used by European companies such as VW and BMW. Tesla has a supercharger connector, which is specific to the Tesla, but there is an adapter that allows the CHAdeMO connector to be used with the Tesla (Tesla, 2015). Herron (2015) has pointed out that the CHAdeMo system was developed in Japan while the SAE CCS was developed to meet SAE standards. All three systems are available in the United States at many locations. There is a need to standardize Level 3 charging (Herron, 2015). Most DC fast chargers are designed to provide rapid direct current charging over a 20–30 min time period with a final charge that is about 3/4 of a full charge. With fast charging 50–70 miles of range are added in 20 min.

There are many places where the EVSE system does not need to accept credit cards or identification cards. In places where the EVSE needs to process credit charges, there are many systems that are able to do this. When a credit card is used, there are often some transaction costs that must be paid. These can be a substantial part of the total bill when the cost of charging is modest.

3.5 Locations for SPCSs

There are three important variations for locations for SPCSs: home, along travel routes, and where drivers stop for an hour or more. Many EV owners will have a charge station at home. This may involve solar panels on a roof or car port. Recently, rapid charging EVSEs have been installed along some interstate highways. Tesla Motors has a network of these in the United States and in Europe. The Tesla high rate EVSE system includes solar panels and batteries for energy storage. Because of the expense associated with rapid charging from the electrical grid, the rapid charging is accomplished using the stored energy in the batteries. There is no charge for Tesla owners to use these charge stations. The third location for SPCSs is where individuals stop for an hour or more, and work sites are the most common of these.

It is becoming increasingly common for work sites to have SPCSs. Other locations where SPCSs may be installed include malls, hotels, gyms, eating establishments, stadiums, parks, churches, and zoos. Service stations may also install SPCSs.

The installation of SPCSs at many locations will help address the range anxiety that affects sales of EVs. If EV owners have a large number of SPCSs at many locations that are available to them, this will allow EVs to be used for more trips. If there were 200 million SPCSs in the United States with an appropriate mix of Level 1, Level 2, and Level 3 SPCSs, the range anxiety issue would be reduced. Many SPCSs that are connected to the grid can be very beneficial even if they are seldom used for EV charging because they are generating clean electricity for the electrical grid.

As EV use grows and demand for SPCSs increases, one variation that is anticipated to become popular is a canopy of solar panels such that entire parking lots are filled with SPCSs. The cost of construction and connection to the grid is less per SPCS when there are many SPCSs. The shaded parking is appreciated by all who park in the lot. Free Level 1 charging can be offered by installing 110 volt receptacles. It is important to be able to use, store, or sell all of the electricity that is generated. When there is a large array of solar panels, there may be opportunities to collect and manage rain water to reduce flooding and make use of the water at a later time.

For homes, garages, and apartment buildings, the solar panels can be mounted to the roof and the charge station equipment can be in the garage or near a parking space along the side of the building. There may be energy storage as well because it can provide electrical power when there is failure in the grid supplied power. This can also be a source of power at night when the solar panels are not producing power. Homes may be the most popular location for SPCSs. Having an EV makes solar panels more attractive for homeowners and having solar panels makes owning an EV more attractive. With time-of-use prices, it may even be best to have excess power produced by the solar panels flow into the grid during the day and then charge the EV with cheaper grid power at night.

3.6 Energy Storage

As the cost of batteries decreases, there will be greater use of energy storage in parking lots with SPCSs and EVSE. Solar energy is available during the day, but not at night. The ability to store electrical energy in batteries has value because it can then be used at a later time when demand is higher. As the sun sets, electrical power needs are often significant (as many people arrive at home after work), and this is a time when stored energy might be used. Stored energy allows the parking lot operator greater flexibility to serve the needs of those who wish to charge their vehicles. With time-of-use prices for electricity, there may be economic benefits of storage that help to pay for the cost of the batteries that are used for storage.

One of the opportunities for energy storage is to take the older batteries from EVs and use them for energy storage in parking lots with SPCSs. The cost of batteries is going down as new developments are commercialized and companies are finding efficiencies. In 2015, battery costs were about $300/kWh (Nykvist and Nilsson, 2015); they are expected to decrease to about $125/kWh by 2022 (USDOE, 2014). As the price of batteries decreases and the percentage of electricity from solar and wind energy generation increases, there will be more battery storage. Time of use prices provide incentives for energy storage in EV batteries and those in parking lots and other locations.

3.7 Business Models for SPCSs

Robinson et al. (2014) describe several business models for SPCSs. The Tesla model is to provide a needed infrastructure with free fast DC charging with the expectation that this will help sales of Tesla cars. As of May 2015, there were more than 400 Tesla Supercharger stations (Richard, 2015). A photo of a Tesla Supercharger station is shown in Figure 3.2. Many employers presently

FIGURE 3.2
This Tesla Supercharger is an example of an SPCS. Solar panels are on the overhead structure with the charging station below. (Photographed by Tesla Press. Tesla Presskit. *Tesla Motors*. n.p., 2015. Web. Jan. 14, 2016. https://teslamotors.app.box.com/pressfiles.)

provide free parking for their employees. It is logical to extend this fringe benefit to free charging of EVs, and some employers have done this. If an employee drives 40 miles to get to work, the cost of charging the batteries at 12 cents per kWh would be about $1.60 for an efficiency of 3 miles/kWh. If the installation of the SPCS system costs $10,000/parking place, and it is used 250 days per year for 20 years, the cost per day is $2.00/day. If 16 kWh are generated and 13.3 are used to charge the vehicle, 2.7 kWh enter the grid. All 16 kWh enter the grid on days when there are no vehicles being charged.

The concept of free EV charging for employees while at work could be justified as an incentive to encourage EV purchases because of the importance of reducing greenhouse gas emissions. For example, the federal government could add SPCSs to its parking lots and provide free charging in the lots.

In many cases the employer can make use of any electricity that flows into the company grid from the SPCSs. As the number of SPCSs increases, there may be a need to make provisions for electricity to flow into the electrical grid that is managed by the electric utilities. Presently, electricity that is generated by solar energy has above average value when time-of-use prices are considered. The lowest prices are at night, and the highest prices are in the late afternoon. On a normal working day, EV owners can plug in when they arrive at work. In the afternoon when most cars are charged, there will be power to flow to the grid from the SPCSs at an above average price if time-of-use prices are used.

For workplace charging, Level 1 and Level 2 charging are sufficient for those workers who park their vehicle for 8 hours or more while they are at work. For those who have a 1-hour commute to work, the vehicle can be nearly completely charged during a full work day with Level 1 charging. For some company vehicles, there may be a need for fast charging if the vehicle is used for business purposes during the day.

In cities, there may be public parking lots where SPCSs can be added or installed when a new parking lot is constructed. There are several options for SPCSs in public parking lots. Where there is free parking, there can be free use of the charge stations as well. This can be paid for through a sales tax or the same source of funds that is used to maintain the free parking. Where there is metered parking, the cost of charging can be recovered from the meter income. In parking garages, the cost of charging can be included in the parking fees. Here the solar panels may be on top of the garage. Another alternative is to allow the local electric utility to construct and operate the SPCSs and collect income from use of the charge station. This may require the utility to work with the regulatory organization that determines their rates to approve a special rate for sale of electricity at SPCSs.

Volta (2015) is a company that sells advertising and provides free EV charging. The idea of using advertising income to help pay for SPCSs can be implemented in many locations and parking lots. The listed price for an SPCS marketed by EcoVantage with Level 2 charging, advertising panels, and LED lights is $17,445 (EcoVantage, 2015).

Envision Solar (2015) markets a solar tree that has solar panels with a tree structure that is 35 feet × 35 feet and tracks the sun. It shades 6–8 parking spaces and generates sufficient electricity for about 700 e-miles each day. It can be installed with or without grid connections, battery storage, and advertising space. Because it tracks the sun, it generates more electricity per unit area than a stationary system.

In some work environments, the employees may need to pay for the cost of the SPCSs. One approach to doing this is to make use of parking permits that allow the user to park in the shade of the SPCS and plug in to the EVSE. In this case, the income from the permits and the electricity that flows into the grid needs to be sufficient to pay for the SPCSs.

3.8 Life Cycle Analysis of SPCSs

Life cycle analysis (LCA) has been used to make comparisons that look at all aspects of a new process or product. Engholm et al. (2013) have completed an LCA of an SPCS. The LCA shows that the SPCS is a very good and appropriate product when greenhouse gas emissions are considered. There is a need for electrical energy to produce the solar panels, but the amount is much smaller than the energy generated by the solar panels over their estimated life. If the electrical power that is needed to produce the solar panels comes from wind or solar energy, then the LCA is even more positive.

When there are new developments that result in modernization or replacement of SPCSs, many parts of the SPCS can be recycled. During the next 30 years, progress in solar energy development should lead to more efficient solar panels that make it appropriate to upgrade the SPCSs.

3.9 Conclusions

One way to have green electricity for EVs is to fill parking lots with SPCSs. There are social, environmental, and economic reasons for installing SPCSs and there are presently many SPCSs in the United States and some other countries. As prices of batteries come down, there will be more battery storage of electricity generated by SPCSs. The most popular locations for SPCSs are homes, places of employment, shopping malls, and along major highways.

References

Ajumni, D. 2015. Jordan plans to build 30 mw solar-powered electric vehicle charging network, *PVBUZZ*, January 10, 2015; http://www.pvbuzz.com/.

EcoVantage. 2015. EcoVantage Catalog, EcoVantage Energy, Inc.; http://www.ecovantageenergy.com/catalog/.

Engholm, A., G. Johansson, and A.A. Persson. 2013. Life Cycle Assessment of Solelia Greentech's Photovoltaic Based Charging Station for Electric Vehicles, Uppsala University, Sweden.

Envision Solar. 2015. Internet Site of Envision Solar International, Inc.; http://www.envisionsolar.com/.

Erickson, L.E., A. Burkey, K.G. Morrissey et al. 2015. Social, economic, technological, and environmental impacts of the development and implementation of solar-powered charge stations, *Environmental Progress and Sustainable Energy* 34: 1808–1813.

Goldin, E., L.E. Erickson, B. Natarajan, G. Brase, and A. Pahwa. 2014. Solar powered charge stations for electric vehicles, *Environmental Progress and Sustainable Energy* 33: 1298–1308.

Herron, D. 2015. EV DC Fast Charging Standards—CHAdeMO, CCS, SAE Combo, Tesla Super Charger, etc., The Long Tail Pipe, August 11, 2015; http://longtailpipe.com/.

Nykvist, B. and M. Nilsson. 2015. Rapidly falling costs of battery packs for electric vehicles, *Nature Climate Change* 5: 329–332.

Richard, M.G. 2015. Tesla's free-to-use Superchargers growing like weeds worldwide. *Treehugger*, May 27, 2015; http://www.treehugger.com/.

Robinson, J., G. Brase, W. Griswold, C. Jackson, and L.E. Erickson. 2014. Business models for solar powered charging stations to develop infrastructure for electric vehicles, *Sustainability* 6: 7358–7387.

STAR. 2015. STAR community rating system, Version 1.2, March 2015; http://www.starcommunities.org/rating-system/.

Tesla. 2015. CHAdeMO Adapters; http://shop.teslamotors.com/products/chademo adapter/.

Tesla Presskit. 2015. *Tesla Motors*, January 14, 2016; https://teslamotors.app.box.com/pressfiles.

U.S. Department of Energy (USDOE). 2013. *Plug-in Electric Vehicle Handbook*, U.S. Department of Energy DOE/GO-102013-3925; http://cleancities.energy.gov/publications.

U.S. Department of Energy (USDOE). 2014. U.S. Department of Energy vehicle technologies office: Plug-in electric vehicles and batteries; http://energy.gov/eere/.

4

Infrastructure for Charging Electric Vehicles

Jessica Robinson and Larry E. Erickson

CONTENTS

Never doubt that a small group of thoughtful, committed citizens can change the world; indeed, it's the only thing that ever has.

Margaret Mead

4.1 Introduction

One of the most important sustainable development challenges for EVs is to establish the infrastructure that is desirable for charging EVs. The infrastructure that is needed includes SPCSs, grid connections, battery storage, transformers, and transmission lines. For many who own an EV, the home installation of an electrical power supply dedicated circuit to charge an EV with a Level 1 or Level 2 electric vehicle supply equipment (EVSE) is the first step. For Level 1 charging, the impact on the local electrical grid is small; however, for Level 2 charging the impact on the local grid is more significant. The power flow for 30 amps with 240 volts for Level 2 is 7.2 kW, which may require caution. There is the potential to overload the local electrical system if the Level 2 EVSE is on while there are other significant loads that are on in the home as well (central air conditioning, electric stove, electric clothes dryer, electric water heater). For Level 2 charging, you should review the capacity of the electrical system in your home and review the capacity of your transformer with your electrical utility. More broadly, if all residences in a neighborhood have Level 2 EVSE systems, this would result in a significant increase in power flow when several of these are on at the same time.

Tesla Motors has started building needed EVSE infrastructure along some major highways to assist EVs with driving long distances. See www.teslamotors.com/supercharger for the latest maps and locations. The EVSEs and SPCSs that Tesla has installed are a big step forward, but much more is needed. More SPCSs and EVSEs need to be installed and made available at locations where people work. To some extent, this is beginning to happen. The U.S. Department of Energy has a workplace charging challenge to encourage greater installation. A number of companies and organizations have signed up to participate in this challenge. Currently all but 7 states have partners participating in this program with 605 workplace-charging locations. The number of partners and charging locations is growing, but greater support is needed (U.S. Department of Energy, 2015).

These initial developments are small compared to what is required to reduce carbon emissions by 80% by 2050. An infrastructure with 200 million SPCSs for 250 million EVs in an environment where 80% of electricity is generated without carbon emissions will require some battery storage in EVs and in stationary batteries. If each SPCS generates 16 kWh/day, 200 million SPCSs would generate about 1 billion MWh/year. This is about the same amount of energy that is needed to power 250 million EVs that drive 12,000 miles/year at 3 miles/kWh. In 2013, the electricity generated in the United States was about 3.7 billion MWh (USEIA, 2015).

There is a need to increase the fraction of electricity that is generated without carbon emissions as well as a need to have a robust infrastructure for EVs. Adding SPCSs to parking lots is beneficial to both the goal of increasing the amount of electricity produced with solar energy and the objective of expanding the infrastructure to charge EVs.

4.2 Controlling Electricity Demand

The current electric utility model primarily practiced is to provide electricity to consumers whenever it is requested. Power plants constantly provide a minimum level of electricity, even at nighttime when demand is extremely low to satisfy the potential need of any consumer. As the number of consumers grows and demand rises, a utility will simply build another power plant to meet increasing need. However, this model is impractical for the widespread adoption of electric vehicle charging. Charge stations can require a significant amount of electricity ranging from a low level of about 2 kW at 120 V for Level 1, a medium level of about 7 kW at 240 V for Level 2, to a high level of about 50 kW at 480 V for Level 3 (DC fast charging). If the majority of consumers charge their vehicles at the same time, for example, immediately after work, a huge strain would be placed on the electric grid at that time. In order to avoid a large load from being placed on the grid, consumers must change their behavior from using electricity whenever they wish to using electricity at off-peak times, when there is not a high demand on the grid. Cost is often an effective mechanism for changing consumer behavior and can be applied to the utility model. Two utility pricing strategies that can be employed nationwide include demand charges and time of use rates. Demand charges are rates charged per kW for the consumer's highest electricity demand for a period of at least 15 min that month. Also, recall from Chapter 1 that time of use rates cause the cost of electricity to vary depending on the time of day or current demand on the electrical grid. The cost of electricity increases when there is a high demand on the grid (such as immediately after work), and the cost of electricity decreases when the demand is low (such as at nighttime). Using these price mechanisms, consumers will become more conscious about the time they charge their vehicles and will be more inclined to do so at off peak times, such as late at night and early in the morning, to save money. Consumers will have an enhanced awareness overall for when they use their electric appliances, such as clothes dryers, air conditioners, stoves, and SPCSs, and will attempt to use them at staggered times. Thus, the demand for charging EVs and for electrical needs in general will be more distributed throughout the day, preventing grid overload and potential blackouts (see Chapter 6 for additional information on handling high demand on the electrical grid).

4.3 Electricity Generation, Transmission, Distribution, and Smart Grids

How is electricity actually delivered to residences and businesses? Electricity generated from power plants is transmitted via power lines and a transformer. The transformer steps up the electricity voltage to 150,000 V–760,000 V to reduce the amount of energy lost due to resistance during transmission (Resistance = Current/Voltage). The electricity travels across the power lines at this high voltage level. Then, before reaching the final destination, the voltage is stepped down again (with another transformer) to a safer level of 120 V–240 V. The electricity is transmitted in the form of alternating current (AC). However, most electronic devices must first convert the electricity from AC to direct current (DC) because they require this electricity form to operate. In contrast, battery storage and solar power systems produce DC. When electricity is transmitted from a power plant to an SPCS or vice versa, the electricity must first be converted to the corresponding form. For example, sending generated power from a solar panel system back to the electrical grid first requires the direct current to be converted to AC using a power inverter, resulting in about 2%–5% energy loss. As the number of residential and commercial solar systems increases, this process will become more commonplace. With the employment of smart grid, DC to AC and AC to DC conversions will happen often as electricity is enabled to travel in multiple directions. In addition to multidirection travel of electricity, a smart grid will allow greater communication between all entities involved in electricity generation and transmission. For example, using a smart grid, excess solar energy generated can be delivered to the grid and used at another location that currently has a greater demand. This prevents the loss of excess solar energy generated and reduces the amount of electricity power plants must generate. It can also create an attractive business opportunity for proprietors by providing a credit or profit if permitted to sell back the excess solar energy generated. Smart grids will play an important role in conjunction with SPCSs. For example, a parking lot may have empty SPCS stalls at some points in the day. Despite having no car charging, the solar panels will continue to generate electricity. Without a smart grid or a battery connected to the SPCSs, the generated electricity would be lost. However, with a smart grid the electricity could be delivered to the grid, providing clean energy elsewhere and potentially providing credits for the SPCS owner (depending on utility regulations).

4.4 Cost and Construction Requirements of SPCSs

The cost of an SPCS varies depending on the level, location, features, vendor, and solar panel cost. Installation costs are higher for charge stations with greater electrical complexity. Level 1 SPCSs are typically the least expensive while Level 3 SPCSs are the most expensive. Location also affects the installation costs, including factors such as the station's distance from the power source, the amount of construction and concrete necessary, and how old the local electrical system is (e.g., an old house versus a new house). Of course, adding features such as radio frequency identification (RFID), networking, and near field communication (NFC) to the charge station will also increase the price. RFID allows only those with an RFID card or access code to use the station and adds optional billing capabilities such as payment using a credit card; networked charge stations allow charge station data to be collected, notifies the user when charging is complete, and pinpoints the station on an EVSE locator map; NFC is a developing technology, which enables users to activate and pay for charging via their smart phones.

Solar powered charge station costs can also vary depending on the vendor for the charge station equipment, solar panels, and solar panel installation. Larger solar array systems will be more expensive than SPCSs with smaller solar array systems. Finally, public stations are typically more expensive than private stations since greater safety precautionary measures are taken and often more features are added. Thus, given the number of variables that affect the installation cost for SPCSs, it is difficult to provide an average price. On top of all this, there is the background variability and the ongoing decreasing cost of the technology. Very generally, though, with costs of solar panels decreasing to below \$1/W, the cost of SPCSs is in the range of \$10,000–\$30,000 per parking space.

Agenbroad and Holland (2014) have reported installed prices for EV charging stations of about \$1200 for a Level 2 home charger, about \$4000 each for five public parking garage stations that have Level 1 and Level 2 capability, and about \$60,000 for a DC fast charger (Level 3). These costs are for the charging station; they do not include the cost of the solar panels and supporting structure.

There are several mounts one could use to install a charge station; these include ceiling mounts, wall mounts, or floor mounts. Ceiling mounts and wall mounts are mainly used for charge stations installed in residential areas, while floor mounts are used for most public outdoor stations. Ceiling mounts can minimize tripping hazards, but reduce headspace and can be an obstacle for cars. Wall mounts may allow for cheap electric installation and do not take up any square footage. Floor mounts usually take up the most space and require cement work.

The Society of Automotive Engineers (SAE) created universal plugs, the J1772 and J2293, to enable EVs to plug-in with any EVSE, assuming the manufacturer complies. This provides EVs greater flexibility, preventing EVs from being limited to only plugging into certain companies' charge stations. As of 2015, only Tesla vehicles have the equipment capable to use Superchargers. However, CEO of Tesla Motors, Elon Musk, has announced his willingness to share this technology with other EV manufacturers, so competing EV models may also be able to use Superchargers in the future.

Ye et al. (2015) have reported a cost of energy of \$0.098/kWh for an integrated SPCS system connected to the grid with time of use prices that encouraged night time EV charging from the grid. Time of use prices for electricity and low interest rates on the capital investment are important factors that impact the economics of their study.

4.5 SPCS Locations

There are at least two reasons why it is important to provide SPCSs in many locations. First, this provides people a choice of where to charge. It provides flexibility by offering people the opportunity to charge, for example, at work or during an event later that night and spreads out demand on the electrical grid. Second, greater location options also increase the chance EV owners can conveniently charge their vehicle at work, while running errands, or at an event rather than going out of their way to find an SPCS. Increasing the availability of SPCSs reduces EV owner frustration and makes EVs more attractive to consumers.

There are a multitude of places SPCSs can be located, including work places, residential areas, shopping centers, theaters, motels and hotels, public parking lots, parking garages, city street parking, universities, sport stadiums, restaurants, arenas, zoos, and rest stops. Arguably, stops off major highways and work places should be the first priorities for providing sufficient charging infrastructure. Drivers often cite range anxiety—fear an EV will lose charge before reaching a charge station—and the inability of taking a road trip as reasons preventing them from purchasing EVs. These obstacles could be resolved if sufficient charging infrastructure was built along major highways. Drivers would feel more confident driving farther distances such as from their residence to the city or taking a road trip. Work places should also be a priority for equipping with charge stations because depending on commute distance some drivers may not be able to make a round trip to and from work on a single charge. For example, if a driver commutes 50 miles to work using an EV with an 80-mile range, but the workplace does not provide charging stations for employees, the driver would not be able to make the drive home on a single charge. The driver could stop to charge off the highway on the way home, assuming sufficient infrastructure was built, but this is an added inconvenience. Most people commute to work five days of the week and must be able to count on their vehicle making the journey. Charging while at work is an important convenience that promotes EV adoption.

A parking lot with SPCSs is being constructed at Las Positas College in Livermore, California with battery energy storage and grid connections. There are projected energy savings of $75,000 per year for this 2.35 MW solar array, which will be operated as a microgrid with 250 kW/1 MWh batteries and Level 2 charge stations. The system will be used to reduce peak power and balance energy loads (Herron, 2015; Imergy, 2015). The 2.35 MW solar array will generate about 55% of the electrical energy needed at the *entire* campus.

Section 4.6 discusses business models for various SPCS locations, building off an earlier publication by Robinson et al. (2014).

4.5.1 Level of Charging Fit for SPCS Locations

Level 1 charging is sufficient for many vehicles at residences, motels, and hotels because EV owners will be parked overnight providing adequate time to fully charge a battery. The charge stations will also be relatively simple to install and operate, and will not have a large impact on the electrical grid. Level 2 charging is needed for overnight charging of EVs with larger batteries (10 h × 7 kW = 70 kWh). The power generated by the solar panels over one parking place is similar to the power delivered with Level 1 charging. In situations where EV owners could spend a various amount of time parked, ranging from 1 hour to several hours, it would be beneficial for locations such as shopping centers, public parking, and university campuses to provide options for both Level 1 and Level 2 charging. For most locations, Level 1 or Level 2 charging is sufficient. Level 3 charge stations are often best suited for locations where EV drivers have a limited amount of time they can (or want to) be off the road, such as at rest stops along major highways. Level 3 charging is the most complex level and requires rapid charging from batteries or the electrical grid.

4.5.2 Long Distance Trips

When traveling on long distance trips with an EV, it is necessary to recharge one's vehicle approximately every 80–200 miles (depending on battery and vehicle type). In order to satisfy this requirement, rest stops along major highways and interstates should be equipped with charge stations. Tesla is accommodating this need by building Superchargers at rest stops and restaurants along major highways. NRG (2015) has been installing fast DC charging stations in the United States in cities with both CCS and CHAdeMO connections for EVs that have the capability to use them. Fastned is a network of SPCSs in the Netherlands that have fast DC charging capability at about 50 kW of power. The construction goal is to have 50 SPCS sites in 2015 and 130 SPCS sites by 2017. The charge stations have both CCS and CHAdeMO connections. The solar panels help give visibility to the stations, there are multiple charging stations at each site, and storage batteries are included in the design to provide buffer and help manage peak demand on the grid (Langezaal, 2015). Locating charge stations at restaurants conveniently allows drivers to eat a meal while simultaneously charging their EV. However, since the average 2015 EV must recharge about every 80 miles (200 miles for Tesla), stopping at restaurants every time to recharge is not realistic for all EVs since the driver may not be hungry that often. Therefore, rest stops should be equipped with SPCSs and additional amenities to occupy the driver as their vehicle recharges. These amenities could include Wi-Fi access, tourist information, a walking path, a television or selection of movies, library, games, and a playground. Families would be entertained as the car recharges and drivers would have the chance to stretch their legs. Lastly, it is worth keeping in mind that the single-charge range of EVs is steadily increasing. The Nissan LEAF's maximum range is now up to 107 miles, and the Tesla Model S 70D has a range of 240 miles. Current projections are that future models of these cars will soon have about twice those ranges.

4.5.3 Potential Future Issues

4.5.3.1 Long Distance Trips and Major Events

As the number of EVs on the road increases (Electric drive sales dashboard, 2015), a potential infrastructure issue could arise on holidays and at major events when trying to satisfy the need of many EVs charging simultaneously. As families travel to relatives' homes for Thanksgiving or Christmas, a flux of vehicles could arrive at rest areas. With insufficient charging infrastructure, a queue of low-charged EVs could quickly form and contribute to charge station traffic. In addition, when a major baseball game, football game, or concert occurs, a large number of fans dispersed throughout the state may flood toward the stadium or arena, each expecting to charge their EV on arrival. Without correct planning, there will not be enough SPCSs available for all EV owners to charge their vehicle and the sudden, massive number of vehicles charging could strain the electrical grid. For major events, transportation officials and event organizers could stress the importance of EV owners to charge along the drive to the stadium or arena and back, rather than relying on charging during the game or concert. Another proactive planning action to avoid this issue could be to build sufficient SPCS infrastructure along major highways to meet the high demand of holidays and major events. This strategy allows the infrequent, large charging demand to always be met and permits any unused solar power generated to be delivered to the grid to earn the proprietor a credit. Stadiums and arenas should also install SPCSs to enable at least a portion of the attendees to charge their vehicles during the event. Since most concerts and some sporting events occur at night, it is necessary for these SPCSs also to have storage capabilities. Battery packs of about 20–35 kWh per parking space could be installed with each charge station to allow for 4–5 hours of Level 2 charging at about 7 kW of power. This is sufficient to recharge most drivers' EVs; those who live locally and those who commute far distances. When the lot is vacant during the day, solar power can charge the stations' batteries in time for events in the evening. When there are no home games or concerts, the battery storage can be used to meet peak power needs or potentially deliver to the grid (depending on utility regulations) earning the proprietor valuable credits or income during an otherwise negative revenue time period.

4.5.3.2 Multi-Unit Residences or Homes without Garages

As the percentage of EVs increase, more consumers who live in multi-unit residences (apartments and condominiums) or homes without garages may own an EV but not be able to install a personal home charge station. This may even prevent these consumers from purchasing an EV.

Presently, there are many steps an EV owner must follow in order to have a charge station installed at their multi-residence complex. For example, a consumer must first get permission from the home owners association and property manager, find a suitable spot for the charge station with access to 240 V electricity, and work out who will pay for the installation, maintenance, and so on. Some states, such as Hawaii and California, have laws protecting consumers' rights to install a charge station (Webb, 2014).

Apartment and condominium owners should consider installing charge stations on-site for their residents. The process can still be complicated in terms of finding suitable parking spots and working with utilities, but it would help retain residents and attract new residents who own EVs. Powertree Services is bringing SPCSs to apartment dwellers in San Francisco by installing solar panels on apartment buildings and parking garages and renting spaces for EVSEs in the garages to allow EV charging at flat rate prices designed to allow travel costs to be about one-third of those for gasoline vehicles. The electricity generated can serve needs within the apartment building if not needed for charging, and the electricity can provide off-grid backup power (Ayre, 2015).

Some EV owners who do not have home garages have chosen to strategically utilize public and workplace charge stations, or to hire an electrician to run a 240 V cable underground and install a protected curbside outlet. As the charge station infrastructure grows, it is more likely a charge station will be located at a public location near a consumer's residence or at a consumer's workplace, reducing the need to charge at home. As more consumers experience similar issues, housing communities may collectively install public charge stations for homeowners to use. In addition, state laws may pass requiring new homes to have 240 V outlets installed.

4.6 SPCS Funding Strategies

When determining how to fund the installation of SPCSs, there are multiple financing models one could employ. Possible sources of funding include federal grants, tax incentives, income from charge station user permits, income from charging an hourly rate, or establishing a partnership with an electric charge station company or utility. The best financing strategies to use mainly depend on who (company, shopping center, residential community, etc.) is installing the SPCSs.

One current example is Volta Charging, a company providing free electric charge stations for businesses (predominantly shopping centers) and allowing consumers to charge their EVs for free (Our story, n.d.). Volta pays for the EVSE installation and electricity cost using advertisements displayed on the charge stations (Our story, n.d.). The ads are primarily for the businesses located in the same shopping center as the charge stations. Businesses are willing to pay for these advertisements because they are directly reaching consumers before they shop (Our story, n.d.). In addition, the retail centers are in favor of the EVSEs because they attract more customers and improve their green image (Our story, n.d.). Although this model will not work well for some other locations, such as rest areas, it could be widely adopted in shopping centers and similar venues. Thus, the use of advertising is also an option to help pay for the cost of providing SPCSs, providing a revenue stream in addition to the value of the electricity that is produced and any consumer cost of charging an EV.

4.7 Electric Bikes, Electric Trucks, and Commercial and Governmental Fleets

4.7.1 Electric Bikes and Electric Motorcycles

When making a quick trip to the grocery store or another nearby errand, electric bikes (e-bikes) are useful vehicles. They have an average range of 10–30 miles depending on the amount of pedaling assistance and can be charged using a home outlet. Ford has unveiled prototypes for two e-bikes that are foldable and have sensors to alert riders of passing cars. The foldable feature enables riders to bring the bikes on public transportation or fit the bike in a car (News Ford Media Center).

Electric scooters (e-scooters) and electric motorcycles (e-motorcycles) are also gaining popularity. Like EVs, they need no gas and little maintenance, and they have ranges of 40–90 miles and over 130 miles, respectively. E-scooters and e-motorcycles can be charged using a home outlet or charged using most public EV charge stations using an adapter. These vehicles have large markets in Asian countries, with China leading in sales. As the number of e-bikes, e-scooters, and e-motorcycles increases, bike racks and public charge spots with the capability to charge these vehicles and EVs will be useful. The charging spots should be mainly located in urban areas at highly trafficked areas such as shopping centers, markets, and work places.

There is significant value to having SPCSs for electric bikes, scooters, and motorcycles because the shade and cover provided by the solar panels protects the vehicles from rain or snow. An infrastructure of parking spaces with SPCSs for these vehicles also encourages purchase and use of these vehicles, which can positively affect the air quality in these highly urbanized locations.

4.7.2 Electric Trucks

There are several strategies that can be adopted to provide sufficient charging infrastructure for electric trucks, and accommodating for their larger battery packs. One option is to employ battery swapping for EV trucks. Trucks could tow a small trailer with sufficient battery energy to travel more than 300 miles. A battery swap company could provide this service by charging EV truck battery packs using solar power and the electrical grid and swapping out EV trucks' low-charged batteries for a fee. When an EV truck is low on charge, it can make a quick stop off the highway to switch out the low-charged battery and replace it with a fully charged-battery. The EV truck could also carry multiple fully charged battery packs and replace the battery pack in use when it loses charge.

Another strategy is to adopt a model similar to that suggested for charging an EV car, which would entail building SPCS infrastructure primarily along major highways. Using this model, EV truck drivers would rest at an EV truck stop while recharging their vehicle. The EV truck stop area could have a restaurant, restrooms, and sell an assortment of food items similar to gas stops (water, sodas, slushies, hotdogs, snack foods, and sweets) as well as convenience goods (lottery tickets, toiletries, tobacco products). In addition, to provide entertainment the truck stop could have Wi-Fi access, a mini-golf course, TV lounge, library, and possibly an area to nap. The truck stop should be equipped with an appropriately sized roof-top solar array to power as much of the truck stop's electricity uses with clean energy as possible to reduce the electric bill and environmental impact. It is key that the truck stop's SPCS infrastructure has sufficient battery storage. Similar to Tesla's Supercharger model, several large battery packs can store excess solar power to reduce the strain on the electrical grid and to help the truck stop avoid utility demand charge fees.

A few electric truck companies are operating already, practicing two different charging strategies. For example, Smith Electric Trucks has fully automatic chargers *on board* the vehicle with the standard connection plugs for the United States, Europe, and Asia (Affordable and Available All-electric, n.d.). Also, Boulder Electric Vehicle has electric trucks that are Vehicle-to-Grid (V2G) capable meaning when plugged into the charging station the EV truck batteries can store electricity and sell it back to the grid at opportune times, earning a profit (Boulder Electric Vehicle, n.d.). Both of these technologies are potential standard features of future EV trucks.

The cost of batteries is steadily decreasing as technological developments are made, but currently the price is still relatively high. In order to make the large solar storage battery packs more affordable for EV truck stops, used EV car batteries can be recycled. Used EV car batteries are less expensive than new EV batteries and typically still have 80% storage capacity left (Used Chevrolet Volt Batteries, 2015). GM has been successfully recycling used EV batteries and installing them in homes, businesses, and utilities to power buildings, serve as back-up power in case of a blackout, and to store excess solar energy if applicable (Used Chevrolet Volt Batteries, 2015). Truck stops could also purchase used batteries to store excess solar energy at a cheaper price. For example, recycled EV batteries cost $200/kWh as opposed to the average $300/kWh for a new battery (Nykvist and Nilsson, 2015). The truck stop could purchase these recycled batteries and charge $.04/kWh for SPCS use. Assuming the SPCS with battery storage is used 300 days of the year, the battery cost would be paid back in 20 years. Since it costs $0.67/mile to drive a gas-powered vehicle on average, the storage cost of $0.04/kWh is very reasonable. As of 2015, there are not yet enough used EV batteries to employ this model of buying recycled EV batteries on a large scale, but it will be a viable nationwide option soon.

4.7.3 Commercial and Governmental Fleets

Commercial and governmental fleets have already started transitioning to EVs. For example, GE has purchased 25,000 EVs for their fleet and fleet customers, and city governments such as Houston and Bay Area cities have transitioned to EVs (Fleet Electrification Roadmap, 2010; Houston Drives Electric; Electric Vehicle Fleet, n.d.). EV fleets promote EV adoption while saving the company or government on maintenance and fuel fees. Because fleets often drive the same route and distance every day, range anxiety for driving EVs is less of an issue (Fleet Electrification Roadmap, 2010). The installation of SPCSs for fleets with fixed routes can be fairly easily accomplished and is very cost effective. In addition, since some commercial or governmental fleets (or specific cars within fleets) often do not travel far distances daily, those EVs could have smaller battery ranges and thus cost less (Fleet Electrification Roadmap, 2010).

Infrastructure needs for fleets vary depending on where the fleets are kept off-duty and miles typically traveled (Fleet Electrification Roadmap, 2010). Fleets where the vehicles travel less than the EV battery range and are stored in a central depot can install SPCSs with storage batteries at their depot location. The EV batteries can recharge while the vehicles are off-duty. Since these EVs are likely charging at night, storage batteries are necessary to capture the solar power generated during the day. Other fleet drivers may park their vehicles at home overnight, often the case for sales or law enforcement vehicles. This would require residential charge stations to be installed as well as several SPCSs at the central depot. Some fleets may not drive predictable routes daily or may travel distances greater than the EV battery range. These fleets must also install SPCSs along streets and highways highly transited and possibly clients' parking lots in addition to the central depot. Infrastructure for these fleets would be more costly given the greater number of SPCSs required, but the fleets would help build the nation's charging infrastructure and possibly qualify for federal government incentives. Since power generated during the day usually has greater value, it may be better to use the power generated during the day at SPCSs for day time activities and peak power needs. Less expensive electricity can be used to charge the fleet EVs at night.

The type of fleet vehicle best suited for a company or government agency depends on the needs of the adopter. Some fleets may need a compact, mid-size, or full-size car while others may require pickup trucks or vans. Currently there are not many EV pickup trucks or EV vans on the market. Nissan Europe is developing a 7-seat electric van for VIP transfers and for hotel, taxi, and private customers. Nissan USA is similarly developing a 5-seat van (Nissan Europe, n.d.; Nissan USA, n.d.). Chrysler is also developing a plug-in hybrid electric minivan (Chrysler vehicles, n.d.), and Via Motors has created an extended range electric pickup truck (Via, n.d.). The small number of EV pickup truck and EV van options does not offer consumers much choice. In addition, these vehicles are in their development stages and may have only recently been introduced. The lack of EVs fit for some fleets' and consumers' needs may be discouraging potential EV purchases. EV pickup trucks and EV vans need to become more available. Several car manufacturers have already shown that the technology for these EVs is available. Expanding to more EV types will appeal to a wider fleet and customer base, increasing EV adoption and leading to greater SPCS infrastructure development.

4.8 Electric Public Transportation

Public transportation such as taxis and buses are one of the main forms of transportation in major cities. It is important that infrastructure for electric public transportation is built to accommodate the transition from gas-powered vehicles and to meet the different needs of public transportation.

Since taxis and buses are almost constantly driving customers to their destination or running a time-scheduled route, they often cannot spare a couple of hours or more during the day to charge their vehicles. Instead of using the same charge stations as personally owned EVs, taxis and buses can use wireless charging pads. Wireless charging pads allow an EV to simply drive over a charging pad, located above or below ground, and immediately begin charging the vehicle without requiring the driver to get out of their car or use any plugs (Plugless, 2015). The EV just needs a vehicle adaptor installed on the bottom of the vehicle. Wireless charging providers also have mobile apps that allow users to locate vacant charging pads; halt charging manually or automatically by choosing to charge based on battery charge, time, or dollars spent; and keep track of monthly usage statistics. If wireless charge pads were installed citywide, a bus or taxi driver could simply pull up to the bus stop or a designated roadside spot while waiting for a new customer and seamlessly recharge their EV. Bus routes are typically the same every day, enabling the bus driver to know when and where the bus will need to charge daily. Many taxis also frequently service locations, such as airports, hotels, convention centers, and tourist spots.

Allowing taxi or city bus drivers to recharge without going off duty for a couple of hours saves the driver time and money. Overnight or whenever the taxi or bus is off-duty, the vehicle can recharge in its fleet parking lot using SPCSs with battery storage. The battery storage is important to capture excess solar power generated during the day in order to charge the vehicles at night. Buses would require a relatively large energy storage battery pack, ranging from about 30–300 kWh, to fully recharge the vehicle and to avoid demand charge fees from utilities. Coupling SPCSs with batteries allows clean energy to recharge a portion of the vehicle, which reduces electrical bills and the strain placed on the electrical grid.

Similarly, school buses follow the same route each morning and evening, every day, so the drivers know the distances the vehicles travel each trip (and there is often some nonoperating time at mid-day). It would be feasible to transition to electric school buses and provide sufficient charging. Electric school buses could use SPCSs installed in the fleet parking lot and charge their vehicle while the children are in school and at night. Depending on the bus route distance, some electric school buses may require wireless charging, which could be installed at various school bus stops.

Infrastructure for both personally owned EVs and electric public transportation should be available in cities to accommodate everyone's needs. In order to accommodate those who drive their personal EVs to the city, sufficient charging infrastructure for personal EVs should also still be present. Given limited city parking availability, parking garages can be equipped with charge stations and rooftop solar. Also, SPCSs can be installed in city parking spots along major roads and be metered with a certain hour time limit to allow vehicle turnover.

4.9 Future of EVs and SPCS Infrastructure

Erickson et al. (2015) report that U.S. PEV sales grew 33.9% from January–June 2013 to January–June 2014. Assuming this annual growth rate remains constant, by 2030 there will be more than 10 million PEVs in the United States. In addition, if the current ratio of 34 PEVs to public charging stations remains constant, there should be over 290,000 public charging stations built by 2030. Maintaining this ratio requires installing 270,000 public charging stations over 15 years or 18,000 public charging stations per year. Ideally, the ratio of 34 PEVs to one public charging station should be reduced even further; but achieving this would require a greater charge station installation rate, assuming the PEV sales annual growth rate remains constant.

Since SPCSs provide a way to generate clean electricity, many parking spaces can be equipped with SPCSs to substantially increase the portion of electricity generated using renewable energy. It is crucial to continue building SPCS infrastructure to meet increasing PEV, BEV, HREV, and EREV demand worldwide. Insufficient infrastructure could lead to queues and frustrated drivers and could perpetuate drivers' range anxiety. In order to effectively transition from gas-powered vehicles to electric vehicles, SPCS infrastructure must be readily available everywhere. Some U.S. and European locations, such as California and major European cities, have greater charge station infrastructure than others. Ideally, consumers will have the option to charge their vehicles where they park 95% of the time. Workplace charging and charging along major highways should become a norm. Leaders in sustainability are working to build the world's charging infrastructure, but progress must continue.

One of the considerations when costs and economics are addressed is the significant reductions in the prices for solar panels and batteries. Prices of both have decreased in the last five years. Because of this, costs from past projects may be higher than those for future projects. Straubel (2015) has shown that the prices for solar panels for generating electricity have followed a model in which $\log Y = A \log X + B$, where Y is the price of solar panel modules in dollars/W of power and X is the cumulative solar PV shipments in units of MW power. An approximate fit of the data gives $A = -0.325$ and $B = 1.67$. That is, the prices have decreased to less than $1/W (Straubel, 2015).

4.10 Conclusions

Infrastructure is arguably the largest barrier for widespread EV adoption. The model for EVSEs must vary from gas station infrastructure and be integrated with frequented locations. Drivers must be able to charge their vehicle simultaneously while running errands, working, or parking at home. This infrastructure must be prevalent not only for private cars, but also fleets and public transportation. Level 2 and Level 3 infrastructure must be built along with wireless charging pads and charging infrastructure for fleets and taxis. Powering this infrastructure with renewable energy, specifically with solar, will generate sustainable energy and reduce the additional strain on the grid. It is crucial to build sufficient SPCS infrastructure to support the transition from gasoline powered vehicles to EVs.

References

Affordable and available all-electric trucks are available today worldwide. (n.d.). Retrieved July 24, 2015, from http://www.smithelectric.com/.

Agenbroad, J. and B. Holland. (2014). Pulling back the veil on EV Charging station costs. RMI Outlet, Rocky Mountain Institute, April 29, 2014; http://blog.rmi.org/.

Ayre, J. (2015). Awesome solar-powered EV-charging station initiative in San Francisco aids apartment dwellers, *Clean Technica*, January 2, 2015: http://cleantechnica.com/.

Boulder Electric Vehicle. (n.d.). Retrieved July 24, 2015, from http://www.boulderev.com/index.php.

Chrysler vehicles. (n.d.). Retrieved July 27, 2015, from http://www.chrysler.com/en/.

Electric drive sales dashboard. (2015, June). Retrieved July 22, 2015, from http://electricdrive.org/index.php?ht=d/sp/i/20952/pid/20952.

Electric Vehicle Fleet National Demonstration Project. (n.d.). Retrieved July 27, 2015, from http://www.scwa.ca.gov/electric-vehicles/.

Erickson, L. E., A. Burkey, K. G. Morrissey et al. (2015). Social, economic, technological, and environmental impacts of the development and implementation of solar-powered charge stations. *Environ. Prog. Sustainable Energy.* 34: 1808–1833. doi:10.1002/ep.12163.

Fleet Electrification Roadmap. (2010, November). Retrieved from http://www.electrificationcoalition.org/sites/default/files/EC-Fleet-Roadmap-screen.pdf.

Herron, D. (2015). Solar panel covered parking lots with charging stations underneath—EV nirvana, *The Long Tail Pipe*, July 3, 2015; http://longtailpipe.com.

Houston Drives Electric. Green Houston Electric Vehicles. Accessed August 7, 2015. http://www.greenhoustontx.gov/ev/.

Imergy. (2015). Chabot—Las Positas Community College District, Imergy Power Systems and Geli Awarded CEC Grant to Provide Energy Storage Technology for Las Positas College Microgrid Project, Imergy Power Systems, February 12, 2015; http://www.imergy.com/.

Langezaal, M. (2015). The fastened freedom plan, *EV Obsession*, October 8, 2015; http://evobsession.com/.

News Ford Media Center. Accessed August 7, 2015. https://media.ford.com/content/fordmedia/fna/us/en/news.html.

Nissan USA. (n.d.). Retrieved July 27, 2015, from http://www.nissanusa.com/.

Nissan Europe. (n.d.). Retrieved July 27, 2015, from http://www.nissan-europe.com/.

NRG. (2015). We make EV driving easy; http://www.nrgevgo.com.

Our story. (n.d.). Retrieved July 22, 2015, from http://voltacharging.com/about.

Plugless. (2015). Meet the Plugless L2; https://www.pluglesspower.com.

Robinson, J., G. Brase, W. Griswold et al. (2014). Business models for solar powered charging stations to develop infrastructure for electric vehicles. *Sustainability* *6*(10): 7358–7387. doi:10.3390/su6107358.

Straubel, J. B. (2015). Energy storage, EVs and the grid, 2015 EIA Conference, Washington, DC, June 15, 2015; http://www.eia.gov/conference/2015/.

U.S. Department of Energy. (2015). EV everywhere workplace charging challenge, Office of Energy Efficiency and Renewable Energy, Washington, DC.

USEIA. (2015). Summary statistics for the United States, 2003–2013, U.S. Energy Information Administrations; http://www.eia.gov/electricity/.

Used Chevrolet Volt Batteries Help Power New IT Building. (2015, June 16). Retrieved July 24, 2015, from http://www.gm.com/content/gmcom/home.html.
Via. (n.d.). Retrieved July 27, 2015, from http://www.viamotors.com/.
Webb, A. 2014. With planning, electric vehicle ownership is accessible to apartment dwellers. Last modified February 4, 2014. Accessed August 7, 2015. http://www.plugincars.com/planning-electric-vehicle-ownership-accessible-apartment-dwellers-129340.html.
Ye, B., J. Jiang, L. Miao et al. 2015. Feasibility study of a solar-powered electric vehicle charging station model. *Energies* 8: 13265–13283.

5

Batteries and Energy Storage

Larry E. Erickson and Jackson Cutsor

CONTENTS

The science of today is the technology of tomorrow.

Edward Teller

5.1 Introduction

Batteries for electric vehicles (EVs) and for energy storage are crucial for the future because energy storage is critical in EVs, and it has the potential to be of great value in smart grids with significant solar and wind power generation. Progress in the development of batteries has changed the world. The cost of batteries for EVs has decreased over the last decade, and this has enabled EVs to become competitive with other alternatives. Energy storage with batteries is now of interest for home owners and businesses that generate electricity with solar panels and it is also being used in power grids to help manage loads and supplies in utility systems. Energy storage is one of the alternatives for the management of solar powered charging stations (SPCSs) in parking lots. Tesla uses batteries for energy storage in its Supercharger network to avoid high power demand charges. Energy storage using batteries has been used for many years for off-grid electrical systems. Since wind and solar energy are generated without any control of wind speeds and solar radiation intensity, energy storage added to the grid provides the ability to generate power at one time and use it at another time. This ability mitigates one of the primary concerns about solar and wind generated energy; maintaining the ability to obtain power whenever it is needed.

5.2 Batteries

Batteries have been common for many years. Lead acid batteries have been used for the electrical systems in vehicles for many years and they were used in the early version of the EV that General Motors introduced but later removed from the market. Batteries have an anode (positive side) and a cathode (negative side); they store chemical energy and make use of oxidation and reduction reactions. The progress with batteries has included increases in energy density (energy/mass and energy/volume) in addition to the significant reduction in cost. Batteries for EVs, just like any other batteries that are used to store electricity as chemical energy to be used later as electrical power, have both power and energy features. The following is a brief summary of these general features. The capacity of the battery is the amount of energy that can be stored in kilowatt hours (kWh) while the power that the battery can deliver has units of kilowatts (kW). The batteries for EVs can be recharged and used many times. The calendar lifetime and the number of times the battery can be recharged are important features of batteries. The battery life is expected to be longer if it is not fully discharged. The depth of discharge is the fraction or percentage of the stored energy that has been used prior to recharging. Batteries provide direct current (DC) while the electrical grid has alternating current (AC). An inverter is needed to change from one form to the other. There is always some loss of energy in going from AC to DC and from DC to the chemical energy in the battery.

A battery management system (BMS) is one of the features of the batteries in an EV, which many other batteries do not have. The BMS keeps the operation of the battery within safe limits and collects data such as temperature and the state of charge. The batteries are arranged in battery packs, which have modules with many cells in each module. The Tesla S has about 7000 battery cells, and the total weight of the battery packs is about 600 kg or 1323 lb (Masson, 2013). The temperature of the batteries is important because the rate of chemical reactions varies with temperature. The power provided by the batteries is less when they are very cold. Very high temperature can damage batteries. Some heat is generated when the batteries are charged and discharged so heat transfer to and from the battery pack is important. The heat that is generated is because the efficiency is less than 100%. Cylindrical batteries have some space between cylinders for air flow and heat transfer. The BMS has a thermal control system for the batteries to keep the temperature of the batteries in the desired range of about 0°C to 30°C. Some thermal control systems use air; others use a liquid. The SPCSs provide shade for the

EVs and this has value in hot climates where there is danger of the battery pack experiencing very high temperatures due to solar radiation that heats the EV. The batteries, BMS, thermal control system, inverter, and charge port are designed to provide the power needed to operate the EV and the energy storage for driving significant distances.

The lifetime of a battery pack depends on the temperatures that it experiences, the depth of discharge, and the charging rate. Level 1 and Level 2 charging with modest depth of discharge are not expected to impact the lifetime of a battery pack significantly. There is greater concern about the life of a battery that is subjected frequently to rapid charging at Level 3 (fast DC charging). The expectation is that the battery system in an EV will last the entire life of the vehicle. The Toyota Prius has many cars that have been in service for more than 10 years. Many of these cars have had only one battery over the life of the vehicle, including a Prius with 530,000 miles (Richards, 2013) that went out of service because of an auto accident. This vehicle was used mostly in California where it was not subjected to temperature extremes.

After the useful life of a battery pack in an EV ends because the capacity has decreased, it may still be useful in a stationary application, such as in a parking lot with SPCSs. After the battery life in the stationary application ends, the materials in the battery can be recycled and used again.

Vatanparvar et al. (2015) have studied the heating, ventilation, and air conditioning (HVAC) system in EVs with the idea of optimizing the operation of the HVAC unit under conditions where control may be needed in order to reduce the power it requires so that power can be used by the BMS and the EV drive system. The proposed controller reduces HVAC power consumption when the electric motor is consuming power significantly because of a hill or acceleration and increases HVAC power use when the electric motor is consuming less or generating. This analysis and control are most valuable when the environmental conditions are extremely warm or cold. The authors expect the life of the battery to be extended by reducing the stress on the battery associated with the power needs and the extreme temperatures.

5.3 Battery Costs

The cost of batteries for EVs has been decreasing because of research, development, and experience. The full cost includes costs for battery cells, modules, and systems. The cost for the complete battery system includes the battery packs, BMS, thermal control system, inverter, and charge port. When reporting information on prices, it is necessary to be specific about what is included. Nykvist and Nilsson (2015) report the cost of battery packs to be about \$300/kWh of storage capacity for the market leaders based on prices in

2014 and early 2015. In a March 2015 report, the U.S. Department of Energy (2015) reports an estimate of $289/kWh of useable energy for a production volume of 100,000 batteries per year. Ayre (2015) has reported estimated costs for Tesla in 2015 are about $250/kWh for the batteries in the 2015 Tesla S. After the new Tesla battery factory in Nevada opens and is operating effectively, the reported estimated values are $88/kWh for battery cells and pack level costs of $38/kWh for a combined cost of $126/kWh (Ayre, 2015). For 70 kWh of battery capacity in a Tesla S at $126/kWh, the battery costs would be of the order of $9000 compared to past estimates of $35,000 for batteries costing $500/kWh. The estimated cost for the GM Bolt batteries is $145/kWh for the battery cells (Cobb, 2015), which is of the order of $200/kWh for the battery packs. These reductions in battery costs may enable GM and Tesla to sell EVs with a range of about 200 miles (320 km) for less than $39,000 (Straubel, 2015). The rapidly falling prices for batteries will reach $200/kWh in 2020 and make EVs cost-competitive within the next 20 years (Climate Council, 2015).

The battery prices given above are for lithium-ion batteries. Lithium-ion batteries are popular because of their energy density and power density. Safety, cost, performance, and life span are other features that are important in selecting batteries, and lithium-ion batteries are competitive in these categories also. The supply of lithium for batteries is estimated to be sufficient so that there will not be shortages of lithium (IRENA, 2015).

The larger battery pack provides greater choices for EV owners. Charging when electricity prices are lower, such as at night, can be arranged when the owner has sufficient range to take advantage of the best time of use (TOU) rates. A 200-mile range is large enough that most drivers will want to stop for other reasons prior to reaching the limit of the range of the vehicle.

5.4 Energy Storage

In addition to batteries in EVs, there is significant battery storage of electricity for a variety of other reasons. In 2014, battery cell sales were about $220 million for utility scale applications (IRENA, 2015). This does not include storage behind the meter by customers who have solar panels and battery storage. China, Germany, Japan, and the United States are the leaders in battery storage (Climate Council, 2015). The future for battery storage in Australia also looks very positive (Climate Council, 2015).

There is a transition that has started related to utility scale electricity generation, storage, and delivery. The smart grid with TOU rates is part of the change. In the past and present, small generators powered by natural gas have been started up and operated when there is a need for more electrical power. Because of lower prices for energy storage in batteries compared to constructing a new natural gas powered generator that would be used only when needed, there is a transition from building these new plants to using battery storage. With greater renewable electricity generation, there is also the challenge of what to do when the utility is producing more power than needed. It is not easy or desirable to reduce production with nuclear and coal generation for short periods of time. Battery storage provides an option for solar and wind generated electricity that is not needed immediately. Battery storage also enables the utility to manage the common daily transients by adding to storage when demand is lower and supplies exceed demand. At the peak demand of the day, power can be delivered from energy storage to help meet the higher demand. With wind and solar, there are transients due to changes in wind velocity and cloud cover that can be managed efficiently with battery storage. Battery storage can replace keeping a small generating plant running to provide a spinning reserve that is commonly used to provide grid stability. The expectation is that the capacity to generate power with renewable sources and battery capacity will continue to increase because the cost of this power and battery storage continues to decrease (IRENA, 2015; Climate Council, 2015). The annual revenue associated with utility grid level battery storage is expected to increase from $220 million in 2014 to $18 billion in 2023 (IRENA, 2015). By 2023, 40% of the estimated battery storage for grid applications is projected for the integration of renewable resources into the grid, 37% for load shifting, and 15% for meeting peak power demands (IRENA, 2015).

Since electric power became available, there has always been a market for electrical systems that are not connected to the grid because connecting to the grid is either very expensive or not possible. Solar panels with battery storage are becoming more competitive for this market. The alternative to solar energy is often diesel fuel, which is much more expensive with current gas prices. The decrease in solar panel and battery prices are resulting in more affordable off-grid electrical systems. This is very important for those who have very limited incomes. In many parts of the world, there are people living in rural areas who do not have electricity. For instance, a large part of Africa does not have electricity in rural homes (Sachs, 2015). Refrigeration, communications, and lights at night are important to quality of life and development. There is value in bringing simple and inexpensive electrical systems to locations where there is no electricity. The decrease in prices will be very helpful to many people who can benefit from having electricity. The current market for batteries for these off-grid applications is in millions of batteries per year (IRENA, 2015).

There are many individuals who are installing solar panels on roofs to generate electricity because they are able to reduce the cost of their power. Some are doing this because they want to support the transition to renewable energy sources. In some cases, battery energy storage is part of the installation with the goal of using storage to allow any excess energy to be used at another time such as in the evening. Solar panels and battery storage allow an owner to have greater reliability because, in the case of grid failure, it is possible to disconnect from the grid and use solar panels and battery storage for power (within the capacity of the system). The market for battery storage by home owners and businesses is expected to grow substantially (Climate Council, 2015; IRENA, 2015).

The price of electric power from the grid varies because of available choices for fuel and costs of the fuel. Island populations without coal and natural gas on the island often have higher electricity costs. Solar energy with battery storage is very competitive with alternatives on many islands (Climate Council, 2015). King Island in Tasmania has reduced its diesel consumption for power generation by 50% because it has transitioned to solar energy with battery storage for a significant fraction of its electricity generation.

Utility demand charges are included in some electric bills with the goal of reducing the peak power use of some customers. The demand charge may be based on the average power level for the 15-min period with highest use during the billing cycle. For example, if a customer charges an EV with Level 2 power while also operating a clothes dryer and electric stove, this may result in the highest use during the monthly billing cycle. Demand charges can be as high as $30/kW, which is $300 for 10 kW of power (Wishart, 2013). Battery storage is being used to reduce demand charges. Tesla uses battery storage at some of its Supercharger stations to avoid significant costs associated with demand charges (Halvorson, 2013). These Tesla Supercharger SPCSs have solar panels over the parking spaces and charge stations, which are connected to both battery storage and to the grid. The fast charging can be at rates as high as 120 kW; however, the BMS manages the charging to avoid overheating and overfilling.

Where a parking lot with SPCSs has a significant number of Level 2 charge stations as well as nighttime charging, battery storage may reduce demand charges significantly and allow the parking lot operator to make greater use of electricity generated by the SPCSs. When there is an event with many in attendance, and the parking lot fills with EVs that wish to be charged, having some stored power can be beneficial to reduce the demand on the grid. There are two considerations; one is the demand charges, the other is the ability of the installation and the grid to handle the load.

5.5 Conclusions

There has been significant progress in the development of batteries for EVs and for energy storage in parking lots and other locations. The reductions in costs have made EVs more popular, and in 2015 the global number of EVs in service went past 1 million. The future for parking lots with SPCSs and battery storage to provide electricity to EVs and the grid is good.

References

Ayre, J. 2015. Tesla gigafactory & battery improvements could cut battery costs 50%, *Clean Technica*, September 21, 2015; http://cleantechnica.com.

Climate Council. 2015. Powerful potential: Battery storage for renewable energy and electric cars, Climate Council of Australia; http://climatecouncil.org.au.

Cobb, J. 2015. Chevy Bolt production confirmed for 2016, *Hybrid Cars*, October 2, 2015; http://www.hybridcars.com.

IRENA. 2015. Battery storage for renewables: Market status and technology outlook, International Renewable Energy Agency; http://www.irena.org.

Masson, L.J. 2013. Why Tesla rules: Huge battery with small cells, *Plug-in Cars*, November 6, 2013; http://www.plugincars.com.

Nykvist, B. and M. Nilsson. 2015. Rapidly falling costs of battery packs for electric vehicles, *Nature Climate Change* 5: 329–332.

Richards, G. 2013. Roadshow: Prius goes 530,000 miles on one battery, *San Jose Mercury News*, February 22, 2013.

Sachs, J. 2015. *The Age of Sustainable Development*, Columbia University Press.

Straubel, J.B. 2015. Energy storage EVs and the grid, 2015 EIA Conference presentation, June 15, 2015.

U.S. Department of Energy. 2015. Energy Storage R&D: FY 2014 Annual Progress Report, Vehicle Technologies Office, Energy Efficiency and Renewable Energy, March 2015; http://www.doe.gov.

Vatanparvar, K. and M.A Al Faruque. 2015. Battery lifetime-aware automotive climate control for electric vehicles, Design Automation Conference, DAC 15, June 7–11, 2015, San Francisco; http://dx.doi.org/10.1145/2744769.2744804.

Wishart, J. 2013. Utility demand charges and electric vehicle supply equipment, *CHARGED: Electric Vehicles Magazine*, October 31, 2013; http://chargedevs.com.

6

Electrical Grid Modernization

Matthew Reynolds, Jackson Cutsor, and Larry E. Erickson

CONTENTS

The value of an idea lies in the using of it.

Thomas A. Edison

6.1 Introduction

The current electrical grid is a distribution grid. Thanks to the innovation and application of alternating current (AC) from Nikola Tesla, the grid can transport electricity from generation locations to remote farms and cities that demand electricity. Principles of electrical generation allow for this to be physically and economically possible. By increasing the voltage, the current is decreased. Power loss is equal to the current squared multiplied by the resistance, $P_{Loss} = I^2 R$, so reducing the current is significant to long distance distribution. Due to electromagnetism and the invention of transformers, electricity can be stepped up or stepped down to either send it long distances or be used for house appliances at 110V. Without stepping up the voltage, too much electricity would be dissipated as heat as it is transported and result in lower end power and accruing safety issues.

For the electrical grid to accommodate an ever-increasing demand of electricity, especially via electric vehicles, it will need to be upgraded, particularly at the local level. The increased load may lead to installation of larger transformers and require more heavy duty distribution lines. In general, local utilities estimate the load and demand from each electric vehicle to be the equivalent of about an average sized home. Adding a few cars distributed evenly throughout a city is not a concern, but with increasing demand and trends of purchasing EVs, subdivisions will need upgrades. People of similar socioeconomic status typically live in the same neighborhood or group of neighborhoods. A wealthier neighborhood might be more likely to add EVs than neighborhoods with lower average incomes. Adding one or two EVs to every household with Level 2 charging in a wealthy neighborhood could effectively increase the demand and load for electricity by 150%.

The future of the electrical grid, termed smart grid, includes multiple directions of electrical travel, net metering, and distributed electrical storage (Burger, 2015). The smart grid increases communications between all entities involved in electrical generation, distribution, and usage. The key flows of communication are the regional patterns of electrical usage for producers and individual statistics for consumers (such as how many kilowatts are currently being used by them and their neighbors). Each of these characteristics opens up a vast array of new business opportunities. Multiple directions of electrical travel means that small producers, like houses or businesses, can send power back to the grid. This system of multiple direction of electrical travel also allows homeowners and businesses to have a reduced power bill.

The current pattern of electricity use has a typical demand curve. The demand is relatively low until people get up and go to work at around 8 am. The demand stays high, then spikes from about 4 pm to 9 pm when people go home and are active prior to going to bed. This creates a valley, a plateau, and a peak. The goal for increasing the efficiency of the grid is to manage supply and demand effectively and economically. Electricity cannot be stored easily and plants have a minimum output. If it is not used, the electricity goes to waste. Electricity is wasted when the valley in the demand graph, or any other point in the graph, is below the generator or generators minimum load. A smart grid aims to minimize these gaps in the minimum output and the electricity being used in real time.

An upgrade to the current power grid has been the subject of considerable recent research and development. This smart grid can provide a method of power and information transfer that will revolutionize the industry. The main feature includes the pairing of an information communication system with the controlled electrical flow. This ensures two important aspects: first that customers can be given new freedoms in their power choices, and second that savings on the utility and power generation market can be made. There are certain limitations to the current power grid that lead to the necessity of this upgrade, including power generation costs, renewables integration, and the customer's role in power management.

In order to reduce the economic and environmental costs, we must find a method to increase the efficiency of the power grid. We can add renewable forms of energy production, thus increasing the power quality of the grid. Due to the variability of most renewable power (wind and solar are dependent on weather conditions), integrating these sources into the current grid may lower the reliability. It is necessary to upgrade to a smart grid that can implement time of use (TOU) prices. The current total power need in the world is about 15 TW, and as our population increases, so will this total (U.S. Department of Energy, 2015). With the current grid, expensive peak power generation units are needed. We need to modernize the electrical grid so that more renewable energies can be considered and time of use can be managed.

One shortcoming of our current grid is the lack of communication. Increasing the customers' knowledge of the power they consume, and providing more information for the utility, is necessary in modernizing the power grid (U.S. Department of Energy, 2015). The smart grid and technologies associated with it will provide the best way to integrate solar powered charge stations (SPCSs) into the grid.

6.2 What Is the Smart Grid?

The smart grid is a cyber-physical system that supports and enables enhanced communication, controllability, and responsiveness of highly distributed customers, power generators, storage units, measurement equipment, and information displays in order to intelligently integrate the actions of all connected participants to deliver electricity efficiently and economically with good power quality and reliability (Camacho et al., 2011; Speer et al., 2015). The benefits of a smart grid include a more efficiently operated electric power system, reduced operational costs, integration of renewable generation, and better management of EV charging through TOU prices.

There are many smart grid technologies and theories that are discussed throughout the literature. The main feature they have in common is pairing a communications system with a decentralized electrical grid. The communications, achieved by Wi-Fi, phone lines, satellite, etc., will work to gather data about the system and power generated in order to: (1) inform the customer of their power usage and TOU prices in order to allow more informed decisions, (2) inform the utility of the power produced by distributed generators to allow for higher efficiency and more reliable pricing, and (3) inform the control operators of the state of electricity in all lines to increase safety and decrease outages. There are many technologies that are considered "smart"; however, we are concerned with the main categories of technology, namely: advanced metering infrastructure (AMI) and smart meters, smart inverters, distribution automation, modern communication systems, and demand response.

6.3 Benefits of the Smart Grid

The major benefits of the smart grid include (Camacho et al., 2011; U.S. Department of Energy, 2015):

1. Better accommodation of renewable generation
2. Greater consumer participation in optimizing electricity use with TOU prices
3. Better power quality and reliability
4. Better optimization of assets and operations to reduce costs
5. Better ability to provide all participants with the information needed for making good decisions
6. More efficient integration of SPCSs and EVs into the power grid
7. Reducing greenhouse gas emissions
8. Better integration of energy storage into the grid

One of the strengths of the smart grid is the involvement of advanced information collection. This can involve many different levels of the power generation system, culminating in the ability to accurately and efficiently monitor production, storage, and distribution of electricity. One major player in this field is advanced metering infrastructure (AMI). This technology involves using smart metering that sends reports on energy usage to the utility. These meters include both those placed at residences and along transmission lines, and have higher resolution times to allow for more representative energy data. Through smart meters and smart appliances, the utility can collect a variety of data including power outage, grid voltage and frequency, and even power quality (Kempener, 2013). Kansas City Water Service Department began the installation of an AMI system and reported that customer service trips were reduced by a third and that the technology had added cost benefits. With the use of advanced "smart" metering, customers can begin to be more informed about their power decisions, leading to a more efficient and better managed electrical grid.

Due to the transient nature of renewable energy, resulting power transients can cause problems for the normal operation of the grid. To approach this problem, a smart inverter is considered. The main function of an inverter is to connect an energy source to the electric grid by delivering power, in the traditional case real power. This can be changed by using an inverter capable of using reactive power which will regulate the voltage at sites where renewable energy is used. Smart inverters are also able to communicate data on grid voltage and frequency, and through proper updating and interacting with operators, power outages can be anticipated and averted more efficiently (Kempener, 2013). These inverters were used in Puerto Rico where their ability to control real and reactive power was valuable in the situation of an island economy where assistance from a surrounding grid is not possible. This technology is one of the next steps toward greater control and security of our electrical grid.

A third component to smart grid technology involves the way in which we operate the electrical grid. Distribution automation (DA) refers to the ability of the grid to respond to the different needs of the grid due to the variability of renewable energy. This relates to photovoltaics as well; when the sun does not shine, DA can continue to control the grid by replacing PV energy with another source. This is much like the current use of resources such as natural gas during peak usage times in order to meet power requirements (Kempener, 2013). We can also see how this relates to the area of demand response, or the way in which electrical loads are reduced during peak usage times. These two areas of smart grid applications have many methods that involve the customer on many levels and this leads to some of the challenges the smart grid poses.

The technologies discussed thus far are in use and show promising effects on controlling and operating the grid. More importantly, as renewable energy becomes a larger player in the distribution of power generation, updated methods of distribution and control become necessary. When renewable energy such as wind and solar exceeds 30%, it is important to have a smart grid with TOU prices and good communication to manage operations (IRENA, 2013). DA and demand response (DR) are features of the smart grid that can be implemented to manage the grid. One goal of TOU prices is to have prices that are in good agreement with the actual costs of operation (IRENA, 2013).

There are two major challenges concerning the smart grid. The first of these is the high cost of some of the technology. This high cost is mainly an initial installation cost, and it is possible to achieve returns on this investment; however, many locations lack the funding to pay for these high initial costs. Research is still being done to lower these costs, which will play a large role in smart grid implementation. The benefits of investments in smart grid technology exceed the costs in most applications (IRENA, 2013). For the United States, the estimated net benefits of adding smart grid technology are $20–25 billion, based on the reduced costs for the utilities to provide electricity (IRENA, 2013). The other challenge is determining the ownership of the new information generated by the smart grid. This data will be quite valuable to many people, and only through experience and policy creation can these new regulations be determined. These challenges will require more research as we progress into implementation of both renewable energy and the smart grid.

6.4 EVs and Solar Powered Charge Stations (SPCSs)

The smart grid provides an opportunity for us to expand our knowledge of the electrical grid that powers our homes and businesses. Now that the smart grid technologies have been defined, we will consider its effect on integration of renewable technologies. EVs are the largest step in electrifying transportation, and they will have a significant effect on the economic and environmental status of the country. The SPCSs provide an easy way to charge batteries for those who own EVs and PHEVs. They provide an incentive to purchase those vehicles knowing it will be possible to charge their car as easy as filling up on gas in a traditional internal combustion engine. The smart grid can only improve these efforts.

The EVs and PHEVs are promising developments in the transportation industry, and especially of interest to the smart grid. As one of the major components of the smart grid is controlled production and distribution of power, EVs provide two key components to the smart grid. The first involves the quantity of electric powered cars on the road. While their numbers are small in comparison to the vehicles with internal combustion engines (ICEs), they continue to grow as publicity on green and sustainable technology gains popularity. The current struggle is that, as the size of the EV fleet increases, the demand for electricity increases significantly. This will be a major consideration for the smart grid and involves the second most important aspect on EVs and the grid: temporary storage. Vehicle to grid technology is being developed that will enable an electric car's battery to become temporary storage and a supply of power to the grid (Mwasilu et al., 2014). By using the technology described previously on the smart grid, we could allow a vehicle to charge the batteries during the evening, and supply power during peak usage time via incentives to the customer. These exciting opportunities lead us to another great opportunity that involves SPCSs that are connected to the grid (Mwasilu et al., 2014).

A major challenge we face as the number of EVs sold increases is the source of electric power. While the vehicle itself is green and sustainable, its ultimate power supply may not be. A solution to this is the use of renewable energy from SPCSs. These facilities allow those with EVs and PHEVs to plug their car in at work or at the store to charge while they perform their daily tasks. This is important because it increases the amount of time that the vehicles could be plugged into the grid. As discussed previously, at SPCSs the vehicles could also be used for storage and supply. The added benefit to SPCSs is that their generation of solar power may be supplied directly to the grid during peak power usage. These applications of smart grid technologies could provide a basis for cleaner energy in the transportation industry (U.S. Department of Energy, 2014).

Advanced metering technology for TOU prices has been developed, but it is not a widespread technology presently. Advanced metering allows electric utilities to monitor the produced and used electricity for all small units, like a house, which are distributed generators and consumers. These smart meters could be utilized for more than just homes. They could be implemented at SPCSs to monitor electrical flow for statistical purposes or to charge consumers for their electricity. Currently the selling of electricity is limited in many states, but with widespread adoption and political pressure that could be changed and create many opportunities for business, competition, and methods to financially support charging stations.

Currently, the electrical grid cannot viably store electricity. This problem is being addressed by various researchers. One prospective solution is electrifying the transportation sector where there will always be cars plugged into the grid at any given point and time of the day. In this case, if demand is high enough, whether that be peak hours or a short power outage, the grid could draw energy stored in the cars' batteries. Obviously, the owners would be notified and paid for their services. This idea gives a great deal of flexibility for electric utilities. Currently, generators have to be run constantly to always provide electricity. The problem is that generators cannot produce just enough electricity during the nighttime to meet the super low demand; generators have a minimum generating capacity. All of the unneeded electricity is left unused. With the new wave of electrified transportation, this unused electricity can be utilized to charge EVs at nighttime, when presumably most owners would be charging their vehicles.

Charging a fleet of cars all at the same time poses many new challenges for the grid. Plugging in a car to Level 2 charging is almost the equivalent of adding another home onto the electrical grid (Mwasilu et al., 2014). Owners of electric vehicles, if it is in their interest, will not plug in their vehicle right when they get home because it is peak demand and that is when electricity costs the most (assuming that the grid has TOU pricing, where the price fluctuates along with the demand). But, if everyone theoretically plugs their car in to charge right when they come home in the middle of peak electricity demand, it worsens peak demand and causes a plethora of consequences. For example, if that demand were to exceed the amount of electricity the grid's current generators could produce, utilities would have to construct a whole new plant just to accommodate for that spike for a few hours a day, resulting in even greater unused capacity at nonpeak demand times. Charging strategies and a smart grid can prevent this from occurring as the use of EVs grows, and similar strategies can be applied to accommodate regular development growth as well.

For the following thought experiment, it is assumed that all cars are some form of EV, either PHEV or BEV. It is also assumed that people have been convinced to do a majority of their charging between the hours of 9 pm and 6 am. The last assumption made is that all cars have a variable battery percentage; that not all cars are nearly depleted. The goal is to be able to have every car charged by 6 am. Without any technology or structure, everyone would plug their car in, with variable states of charge, right before they turn in for the night, presumably between the hours of 9 pm and 11 pm. This creates a drastic peak in demand that slowly dissipates as cars get fully charged. The vehicles that only need to be charged a few percentages quickly finish whereas the cars that are almost depleted could take all night on Level 1 charging. It is projected that some home charging will be Level 2 and some will be Level 1. This has the potential to create problems, but also gives more possible ways to avoid creating a new spike in demand and charging all of the cars before the morning.

Ideally, with smart grid technology and communication between the grid manager and the users, everyone can get their desired outcome in this situation. The problem of creating a new spike can be avoided, nighttime electricity prices can stay low, everyone can have their car charged, and society can reap the benefits of having an electrified transportation sector all at the same time. Each car could be plugged into a charger, but not necessarily start charging until it has been instructed to do so with a smart meter or grid manager. For example, cars with charging times of less than an hour could be lined up in a queue to charge. When one is finished, the next one begins. If done correctly, they will all be charged by the morning and no spike in demand would be created. This type of management can be expanded on a much larger scale, although the optimization is more complex with different levels of charging, different geographical locations, and varying percentages needing to be charged for different battery capacities.

Utilities must have coordinated smart charging programs because, with fluctuations in demand and an assumed surge in certain geographical areas in EV usage, the grid will certainly have the potential to be overloaded. This approach of managing within parameters when cars are actually charging works only if the car is plugged in when the consumer claims and if every charging station has the technological capabilities to interact with the grid. This dictation also comes with a level of responsibility: customers will expect their car to be charged when they wake up in the morning. Thus, the success rate in a smart charging program such as this needs to be extremely high in order to continue to have participants and thereby avoid the spike in demand.

Consumers could provide conditions for charging their car. For example, consumers could specify that they would like their car to be charged to 75% at a minimum, they could enter a range of times that their car will be available for charging, or that they only want to charge the car if electricity is below a certain price. This creates flexibility in accommodating all the cars that need to be charged overnight. Consumers also add extra problems because car usage changes throughout the week and people forget to plug their cars in. With data records and time, these weekly, monthly, and yearly changes can be predicted and adjusted for, but it adds to the complexity of charging cars.

One perk of the electrification of transportation is that prices of electricity are fairly inelastic. That is, prices are stable for relatively long periods of time. A smart grid with TOU prices is beneficial to EV owners and SPCS owners.

6.5 Conclusions

The smart grid is very important because it enables better integration of power from SPCSs into the grid. The smart grid can allow utilities to have TOU prices that discourage EV charging at times of peak demand and encourage charging when power is plentiful. The TOU prices make SPCSs more valuable and they decrease the cost of charging for EV owners who charge when electricity prices are low. Because the smart grid allows the integration of more wind and solar energy into the grid and a greater number of EVs connected to the grid, greenhouse gas emissions should decrease substantially after smart grid installations are complete. The smart grid, renewable distributed power generation, and EVs are mutually beneficial. The smart grid has the capability to provide effective operations with battery storage integrated into the system to manage power generation and distribution with large quantities of renewable facilities. Because there is some energy loss associated with transfer to and from battery storage, it is more efficient to have customers respond to TOU prices and use the electricity when it is generated.

The smart grid enhances real-time information and control, which enables better decision making by customers and system operators. Improved sensing and measurement systems enable faster responses, which improves operations and reliability. Better decision support systems and control allow automation that can improve efficiency and grid management. With TOU prices and customer participation, peak power demands can be reduced and electricity use can be shifted to be more in line with renewable power generation. The smart grid is beneficial to utilities because it is more intelligent and resilient, with sensing capabilities for overloads and natural impacts, and automated response capabilities. The smart grid is beneficial to customers because they benefit from the improved efficiency, greater reliability, and opportunity to reduce their cost of electricity by responding to TOU prices.

References

Burger, A. 2015. An Industry First, Distributed Energy Storage System Dispatches Electricity to California Grid. Renewableenergyworld.com.

Camacho, E. F., T. Samad, M. Garcia-Sanz, and I. Hiskens. 2011. *Control for Renewable Energy and Smart Grids*. IEEE. Control System Society; http://www.ieeecss.org.

IRENA. 2013. Smart Grids and Renewables, International Renewable Energy Agency, November, 2013; http://www.irena.org.

Kempener, R., P. Komor, and A. Hoke. 2013. *Smart Grids and Renewables: A Guide for Effective Deployment*. International Renewable Energy Agency.

Mwasilu, F., J. J. Justo, E.-K. Kim et al. 2014. *Electric Vehicles and Smart Grid Interaction: A Review on Vehicle to Grid and Renewable Energy Sources Integration*. Elsevier.

Office of Electricity Delivery and Energy Reliability. *Smart Grid*. 2015.

Speer, B., M. Miller, W. Schaffer et al. 2015. *The Role of Smart Grids in Integrating Renewable Energy*. National Renewable Energy Laboratory, NREL/TP-6AZ0-63919; http://www.nrel.gov.

U.S. Department of Energy. 2014. Evaluating Electric Vehicle Charging Impacts and Customer Charging Behaviors—Experiences from Six Smart Grid Investment Grant Projects. Smartgrid.gov.

U.S. Department of Energy. 2015. *The Smart Grid: An Introduction*; energy.gov.

7

Distributed Renewable Energy Generation

Larry E. Erickson, Jackson Cutsor, and Jessica Robinson

CONTENTS

The 'clean energy' challenge deserves a commitment akin to the Manhattan project or the Apollo moon landing.

Martin Rees

7.1 Introduction

Distributed electrical energy generation has a number of advantages. These include (U.S. DOE, 2007; Johansson et. al., 2012):

1. Increased reliability of the electrical system
2. Reduction of peak power requirements
3. Improvements in power quality
4. Reductions in land-use impacts and acquisition costs for rights-of-way
5. Improvements in infrastructure resilience
6. Reductions in transmission requirements and costs
7. Low cost option in remote environments
8. Greenhouse gas emissions are small for distributed solar and wind energy

Solar and wind energy installed capacity for electricity generation is now growing rapidly because of improved economics and the importance of reducing greenhouse gas emissions. Solar photovoltaic (PV) has become commercially competitive in the United States, and 4.1 GW of new solar deployments have been reported for the first 9 months of 2015 (GTM, 2015). Wind power additions of 4854 MW have been reported for 2014 (Wiser and Bollinger, 2015). Since batteries in EVs can be charged when the price of electricity is low, EVs can help manage the supply and demand balance as the magnitude of renewable electric power generation increases. For EVs, the renewable power is important because of the reduction in greenhouse gas emissions that are desired by EV owners.

In the first 9 months of 2015, 30% of all new U.S. electric generating capacity that came online was from solar installations (GTM, 2015). In Germany, electricity generated from solar energy was about 7% in 2015 (Brown et al., 2015). In the United States, wind power represented 24% of the new electric generating capacity in 2014 (Wiser and Bollinger, 2015). At the end of 2014, U.S. wind capacity was approximately 66 GW (Brown et al., 2015; Wiser and Bollinger, 2015). In December 2013, wind supplied 62% of the electricity in Denmark and 28% of the electricity in Ireland (Brown et al., 2015). Germany plans to produce 80% of its electricity from renewable sources, primarily wind and solar by 2050.

The focus of this chapter will be on solar PV, wind energy, and battery storage. These topics are important and they relate to the infrastructure that is needed to advance the electrification of transportation. One of the options for solar PV is solar powered charge stations (SPCSs) in parking lots.

7.2 Solar Photovoltaic (PV)

Solar energy can be harvested and converted to electrical power using solar cells (PV) and thin films. Solar panels have been added to the roof of many homes and commercial establishments. There are many parking lots with SPCSs and others that have solar PV but no charge stations. In most cases, there is a connection to the power grid.

7.2.1 Basic Principles

A direct current is generated when a PV cell absorbs light or photons and electrons in the solar panel are excited into a conduction band due to the absorption of photons. The solar panel is constructed to have a voltage and current when the solar radiation reaches the solar panel. The photons have a variety of energy levels that are associated with the different wavelengths of light that make up the solar spectrum. The solar panels are designed to convert a portion of the spectrum to electric power. The semiconductors in solar panels have a threshold energy above which photons can be absorbed and electricity is produced. Photons with lower energy and higher wavelengths than the threshold energy do not produce any electricity. The design of solar panels includes semiconductor junctions that have built-in electric fields (Johansson et al., 2012).

The maximum intensity of sunlight is about 1 kW per square meter. The efficiency of solar PV is defined as the maximum power output of the solar panel divided by the power input (light). Present commercial solar panels have efficiencies that range from about 10–20% (Eco Experts, 2015). There are ongoing efforts to improve efficiency as well as efforts to reduce the price of solar panels. SolarCity plans to begin production of a new solar panel that has an efficiency of about 22% and Panasonic has reported efficiencies of 22.5% on their commercial size prototype solar panels that they can mass produce (Hanley, 2015; Wesoff, 2015). The target cost for the SolarCity panels is \$0.55/W. When solar panels are purchased, both cost and efficiency are considered.

The electric power produced by solar panels is direct current (DC), and in many cases an inverter is used to convert it to alternating current (AC) with a frequency of 60 Hz so it can flow into the grid or to an EV. Since the inverter is only about 95–98% efficient, the power that enters the grid is less than that produced by the solar panels.

The performance ratio (PR) of a PV system is defined as the AC system efficiency divided by the standard test conditions (STC) module efficiency (Johansson et al., 2012). Values of 0.70–0.85 are common for PR because of the following:

1. Light intensities, light spectra, and angles of incidence that deviate from standard conditions
2. Module mismatch, cabling, and inverter losses
3. Cell module operating temperatures
4. Debris and dust on the surface of the solar panels

The PR values are often determined for a year.

While dust and debris can reduce the performance of solar panels, natural washing with rainfall is the common method to remove dust and debris. In most cases, the cost of labor to wash the panels exceeds the value of the increased electricity that is produced.

The capacity factor (CF) for production of electricity with solar energy is defined as $CF = (N \times PR)/8760$ where 8760 is the number of hours in one year and N is equal to the estimated full sun hours in a year; that is, the annual insolation in kWh/yr divided by the intensity of full sun for the same area. The area is usually assumed as 1 m^2. The global range of values for CF is 0.08–0.21 for systems without sun tracking (Johansson et al., 2012).

One of the benefits of distributed solar energy is the impact on the distribution grid. Homeowners with solar panels on their property have small distances from generation to use. When SPCSs have a vehicle with a battery charging, the production and use may be mostly at the same location. These benefits are greatest when the capacity of the solar panels is about equal to the electricity that is needed at that location.

7.2.2 Economics of Solar Energy Generation

The average prices for new installations of solar PV systems has been reported by GTM (2015) to be \$3.55/W for residential rooftop, \$2.07/W for nonresidential rooftop, and \$1.38 for utility fixed tilt. These values are based on the name plate estimated direct current power in Watts. These values vary depending on the cost of connecting to the grid, local labor costs, and permitting costs. The electricity generated in one year per Watt of rated capacity would be N × PR. For Manhattan, Kansas with an N value of 4.57 × 365 = 1668 and a PR value of 0.80, we have 1668 × 0.8 = 1334 Watt hours produced annually per Watt of capacity. For a 30-year lifetime of the solar panels and a cost of \$1.38/W of capacity, the average cost per kWh is \$0.035 based only on the installed cost.

7.3 Wind Energy

Wind energy is being installed on a commercial scale in many locations with individual wind turbines that are 1.5–5.0 MW with a diameter of 60–125 m. Wind farms with 50–200 turbines and 25–800 MW capacity are scattered around the world. Denmark is one of the leading countries in Europe; China has installed many wind farms, and is the leader in Asia in wind capacity. Constraints in transporting large blades, towers, and nacelles over land are limiting the size of wind installations on land. The cost and availability of cranes to erect and install the parts are important limitations as well. The wind velocities and qualities at higher elevations above the land surface provide the incentives for towers that are 50–120 m in height (Brown et al., 2015; Johansson et al., 2102). The cost per unit of power produced has decreased as the size and capacity of the wind turbines has increased.

Wind turbines produce mechanical energy that turns a generator that converts mechanical energy to DC electrical power, which is converted to AC power by an inverter. The electricity flows through power lines to a substation where it is connected to the local electrical grid. Transmission lines are needed to carry the power from the substation to where it is consumed. New installations have been limited by the lack of transmission capacity.

Offshore wind turbine technology has advanced in the last 10 years and there are installations near Denmark, United Kingdom, and China (Brown et. al., 2015). In shallow water, land-based turbine designs have been used with modifications to address corrosion issues and wave forces. Foundation designs include monopile, multipile, and gravity base; gravity bases are filled with ballast and carefully located at the site (Johansson et al., 2012).

Brown et al. (2015) indicate that the shallow coastal areas on the east coast of the United States have the potential for 530,000 MW of wind generation capacity. This has the potential to be a major source of electrical power on the east coast.

7.4 Energy Storage

Energy storage with batteries has been an important part of electrical systems for many years, especially in small systems and microgrids that are not connected to a large central grid. With the growth of wind and solar generation of electrical power, energy storage in batteries is becoming more important (Johansson et al., 2012). The reduced cost of batteries is making battery storage a more important option (Nykvist and Nilsson, 2015). With wind and solar generation, energy storage in stationary batteries and in EVs can help provide grid stability and resilience. The cost of battery storage can be compared to the cost of standby natural gas generating capacity. Both may have commercial value as wind and solar grow to provide more than 50% of the electrical power that is produced.

Batteries that are used for stationary storage do not have the same weight and volume restrictions as those used for transportation. Lead acid batteries have been used for stationary storage for many years (IRENA, 2015). The concept of using batteries from EVs for stationary storage after they have lost some of their capacity has been considered, but their availability is limited.

Real time prices of electricity are important because they provide the financial incentives to store energy in batteries when supply exceeds demand and to use the stored energy when demand is high. EV owners can help the utilities by charging their vehicles when prices are low and not charging them at peak demand times when prices are highest. There is no cost to the utility associated with battery storage in EVs.

One important aspect of using EVs to help balance supply and demand of electrical power with a smart grid is the fact that the conversion of grid AC to battery storage is necessary to power the EV. Battery storage is relatively efficient, but there is some loss of energy when grid AC is converted to chemical energy in the battery. When the stored energy is transformed back to grid AC, there is some loss of energy as well. An efficiency of 90% recovery of the grid AC sent to battery storage and then returned to the grid may be possible. When battery storage is compared to the alternative of having real time prices for electricity that encourage customers to shift their use of electricity to balance supply and demand, it is clear that more energy reaches the customer when time-of-use by the consumer is shifted. The smart grid with real time prices will become more important as solar and wind grow in percentage of electric power generation.

7.5 Conclusions

The significant increase in new installations of wind and solar power generation facilities is happening because of improved efficiencies and economics. The transition to electricity generated with renewable energy has made great progress, and it will continue to accelerate. Two associated benefits are reduced greenhouse gas emissions and better air quality.

Because of reductions in the cost of battery storage, the market for storage batteries in parking lots with SPCSs and other locations such as homes with solar panels is growing rapidly. There are significant benefits to renewable energy management that EVs provide because their batteries can be charged when electricity is plentiful.

References

Brown, L.R., J. Larsen, J.M. Roney, and E.A. Adams. 2015. *The Great Transition*, W.W. Norton, New York.

Eco Experts. 2015. Which solar panels are most efficient, The Eco Experts, London; http://www.theecoexperts.co.uk/.

GTM. 2015. U.S. Solar Market Insight Report, Q3, Executive Summary, GTM Research, 2015; http://www.greentechmedia.com/.

Hanley, S. 2015. Panasonic quickly beats SolarCity's solar module efficiency record, *Clean Technica*, October 9, 2015; http://www.cleantechnica.com/.

IRENA. 2015. Battery storage for renewables: Market status and technology outlook, International Renewable Energy Agency; http://www.irena.org/.

Johansson, T.B., A. Patwardhan, N. Nakicenovic, and L. Gomez-Echeverri, Eds. 2012. *Global Energy Assessment*, Cambridge University Press.

Nykvist, B. and M. Nilsson. 2015. Rapidly falling costs for battery packs for electric vehicles, *Nature Climate Change* 5: 329–332.

U.S. DOE. 2007. The potential benefits of distributed generation and rate-related issues that may impede their expansion, U.S. Department of Energy, February, 2007; http://www.doe.gov/.

Wesoff, E. 2015. World's most efficient rooftop solar panel revisited, Greentech Media, October 13, 2015; http://www.greentechmedia.com/.

Wiser, R. and M. Bolinger. 2015. 2014 wind technologies market report, U.S. Department of Energy; http://www.osti.gov/bridge/.

8

Urban Air Quality

Andrey Znamensky, Ronaldo Maghirang, and Larry E. Erickson

CONTENTS

Water and air, the two essential fluids on which all life depends, have become global garbage cans.

Jacques-Yves Cousteau

8.1 Background

Air pollution is defined as the presence in the atmosphere of particulates or gases that can cause some harmful effects to humans, animals, plants, and materials. Scientific data have revealed connections between polluting sources, degrees of exposure, and health risks (Lim et al., 2012). Premature death due to air pollution exposure (indoor and outdoor) currently claims an estimated 7 million human lives worldwide annually, accounting for 12% of the total global deaths per year (estimates from the World Health Organization [WHO], 2014). The estimate for ambient (outdoor) air pollution is 3.7 million in 2012 (WHO, 2015a). In the United States, 12,000–43,000 people died prematurely in 2000 from air quality issues, of which the majority source was passenger car related (3900–12,000 of the total) (Wadud and Waltz, 2011). The vast majority are cardiovascular-related: from strokes, ischemic heart disease, cancer; and also acute respiratory infections and chronic obstructive pulmonary diseases (COPD). Millions of lost work and school days are also the result of high air pollution levels (USEPA, 2015a).

Historical trends and problem areas have been identified, and regulatory standards implemented to address air pollution issues. Based on these trends, and with the passing of the Clean Air Act by the U.S. Congress in 1963, it has been demonstrated over several decades that the country's economic welfare and growth can improve while at the same time decreasing polluting emissions and deaths therefrom. In the last 30 years, U.S. GDP has grown 145% while the aggregate emissions of the criteria pollutants (i.e., particulate matter, ground-level ozone, carbon monoxide, sulfur oxides, nitrogen oxides, and lead) have decreased by 62% (USEPA, 2015b).

Despite the enactment of increasingly stringent regulations and technological innovations in emissions reduction, there is still a need for more sustainable solutions. One of the largest categories of pollution contributors is light-duty mobile sources. According to the EPA, vehicles are responsible for nearly one half of smog-forming volatile organic compounds (VOCs), more than half of the nitrogen oxide (NO_x) emissions, and about half of the toxic air pollutant emissions in the United States (USEPA, 2008). In addition, the transportation sector accounts for approximately 28% of greenhouse gas emissions (Erickson et al., 2015).

Pollution from vehicles is especially problematic in urban areas, where the concentrations of pollutants can build up to dangerous levels, so it is here that alternatives to conventional transportation are particularly needed. Electric vehicles (EVs) present an increasingly promising solution, especially if powered by electricity generated with renewable sources, including solar and wind. EVs provide a variety of environmental, social, and economic benefits. Few factors have greater impact on human health than air pollution, and with a concerted effort to electrify transportation, it is possible to improve ambient air quality and reduce greenhouse gas emissions.

8.2 Ambient Air Quality Standards and Regulations

In an effort to address air pollution issues, the U.S. federal government has enacted several major pieces of legislation, including the Air Pollution Control Act of 1955, Clean Air Act of 1963, Motor Vehicle Air Pollution Control Act of 1965, Air Quality Act of 1967, Clean Air Act Amendments of 1970, Clean Air Act Amendments of 1977, and Clean Air Act Amendments of 1990. The Air Pollution Control Act of 1955, which represents the federal government's first major effort to address air pollution, provided funds only for federal research and training. These efforts were enhanced through a series of further federal legislations, including the Clean Air Act of 1963, which established federal authority to address interstate air pollution, and the Clean Air Act Amendments of 1970, which represents a landmark event in air pollution. The Clean Air Act Amendments of 1970 established new standards, including the National Ambient Air Quality Standards (NAAQS), New Source Performance Standards (NSPS), automotive emission standards, and motor vehicle emissions inspection and maintenance program. Another landmark event in 1970 was the creation of the U.S. Environmental Protection Agency (U.S. EPA). Together, they helped establish the basic framework of air quality management systems in the United States. The framework was further developed through the Clean Air Act Amendments of 1977 and 1990. Various programs, including air quality and emission protections, ozone protection, prevention of deterioration of air quality, and identification of areas that were not in attainment of the standards, have been implemented.

These legislated emission controls required all mobile sources, including conventional vehicles, motorcycles, and aircraft, to abide by these standards or face fines. Further amendments also addressed emissions from industry, such as coal-fired power plants, requiring stricter monitoring practices, and that facilities in violation of the new standards either reduce emissions or be shut down.

In addition to direct controls on industry, the U.S. EPA also required that urban areas in the United States not exceed certain overall levels of pollutants, which became known as the National Ambient Air Quality Standards (NAAQS, 2012). They specify maximum concentrations of various pollutants in a given area for a given monitoring timespan for six pollutants (see Table 8.1). Of the six pollutants, nitrogen dioxide (NO_2), sulfur dioxide (SO_2), ozone (O_3), and particulates with a size less than 2.5 μm ($PM_{2.5}$) are most important for air quality associated with mobile sources. With current vehicles and the use of fuels without lead, CO and lead in urban air have a smaller impact on health. These pollutants have effects ranging from health impacts, particularly those of sensitive groups, to

TABLE 8.1

Air Quality WHO Guidelines and U.S. National Ambient Air Quality Standards[a]

Pollutant	NO_2 (ppm)	SO_2 (ppb)	Ozone (ppb)	$PM_{2.5}$ (μg/m^3)
WHO	0.021 (annual mean)	7.6 (24-hour mean)	51 (8-hour mean)	10 (annual mean)
	0.106 (1-hour mean)	191 (10-min mean)		25 (24-hour mean)
NAAQS	0.10 (1-hour mean)	75 (1-hour mean)	70 (8-hour mean)	12 (annual mean)
				35 (24-hour mean)

Source: WHO. 2015a. Ambient (Outdoor) Air Quality and Health, Fact Sheet No. 313; http://www.who.int/mediacentre/factsheets/fs313/en/; USEPA. 2015c. National Ambient Air Quality Standards (NAAQS); http://www3.epa.gov/ttn/naaqs/criteria.html; USEPA. 2015d. National Ambient Air Quality Standards for Ozone; http://wwws.epa.gov/airquality/ozonepollution/actions.html.

[a] There are NAAQS for carbon monoxide, lead, and PM_{10} in addition to the values shown here.

decreased visibility, to damage to animals, crops, vegetation, and buildings. The World Health Organization (WHO) guidelines are also shown in Table 8.1.

Air pollution is a significant problem in urban areas, where population density is greater and the number of diesel-powered vehicles and poor enforcement of standards are major contributors to air pollution and health problems. Especially concerning is the situation in developing countries, where standards concerning vehicle emissions, fuel quality, and reduction technologies are lax or poorly enforced. This lack of enforcement in developing countries may be due to education, lack of resources, or the lack of political power to enforce existing regulations (Apte et al., 2015).

By decreasing air pollution levels, countries can reduce the burdens of multiple diseases, including strokes, heart disease, lung cancers, and both chronic and acute respiratory diseases, including asthma. In cities such as Delhi, India, and Cairo, Egypt, tens of thousands of people die each year due to air pollution-related diseases. However, poor air quality is a global environmental health problem affecting urban populations in both developed and developing countries alike. In the United States, in particular, the state of California has seven of the country's most polluted cities, and also the strictest air quality regulations (American Lung Association, 2015).

It is understood that vehicle emissions significantly contribute to ambient air pollution concentrations, and that these air quality standards and regulations have resulted in major emission reductions and improvements in ambient air quality. Air quality in California has improved in the last decade due to legislation and initiatives related to battery electric vehicles, which generate no emissions. California is a state where significant progress has been made as a result of efforts to reduce emissions from mobile sources. Initiatives have been undertaken to improve air quality, such as tax incentives for electric vehicles, the introduction of time of use (TOU) rates for electricity, carpool lanes, and issuing extra tax credits for electric vehicle purchases (California, 2014; Drive Clean, 2015; Lurmann et al., 2015).

8.3 National Air Quality Trends

In order to recognize the importance of expanding research and education related to solar power charging stations (SPCSs) and electric vehicles (EVs), it is important to recognize how the concentrations of the various pollutants have changed over time and where they are coming from.

Sources of pollution include primary pollutants such as those in mobile emissions, windblown dust, commercial products, agricultural sources, and emissions from industrial processes and coal-fired power plants. Of the criteria pollutants in Table 8.1, conventional vehicles have historically been large contributors. Internal combustion engine (ICE) vehicles emit all of the primary air pollutants and contribute heavily to the formation of ozone. For example, Table 8.2 shows the percentage of NO_x emissions by source, with a majority coming from mobile sources (USEPA, 2015e). Some compounds, which have primarily mobile sources—namely NO_x and VOCs, can react in the atmosphere, in the presence of light and heat, to form secondary pollutants such as ground-level ozone. In addition to contributing to climate change, ozone is known to be a human health hazard, irritating the respiratory system and reducing lung function. Primarily found in urban environments, ground level ozone is responsible for around 22,000 deaths each year in the EU (WHO, 2008).

Over the last three decades, aggregate emissions of the criteria pollutants have decreased across the United States, largely due to technological improvements such as catalytic converters and initiatives making them widespread (e.g., the previously mentioned regulatory policies and stricter fuel standards). Still, because of the growing world population and global economy, conventional vehicle sales and miles traveled have continued to increase. According to the World Health Organization (WHO), an estimated 3.7 million people die annually worldwide from strokes, lung cancer, and ischemic heart disease related to outdoor emissions from mobile and industrial sources, close to 6% of all deaths worldwide (WHO, 2014, 2015a).

Countries such as Norway, which have higher market shares of EVs, have concurrently made progress in reducing the concentrations of pollutants in the atmosphere (Norway, 2012, 2014). In 2015, electric vehicles (PHEVs, PEVs, BEVs) comprise only 1% of the total vehicle market volume in the United States. This low volume is especially problematic in urban areas that are prone, either by local geographical, meteorological conditions, or wind patterns, to accumulate pollutants to increasingly dangerous concentrations.

TABLE 8.2

Percentage of Emissions of Nitrogen Oxides by Source Sector in the United States in 2011

Source	Percentage
Mobile	57.9
Fuel combustion	23.9
Industrial processes	8.46
Biogenics	6.60
Fires	2.56
Miscellaneous	0.56
Solvents	0.02

Source: USEPA. 2015e. Air Emission Sources; http://www3.epa.gov/air/emissions/index.htm.

8.4 Environmental and Economic Impacts of EVs and SPCSs

All around the world, major urban areas have become congested with ICE vehicles, resulting in severe air quality problems. Recent efforts have studied the effects of various strategies to reduce motor vehicle emissions and their subsequent impacts (Soret et al., 2014). Electrification of transportation is one of the most promising strategies for improving urban air quality, particularly if combined with renewable electricity generation. It carries with it impacts in social, economic, and environmental areas (Erickson et al., 2015). Adoption of EVs may substantially reduce emissions of greenhouse gases, improve regional air quality, increase energy security, and take advantage of inexpensive solar power (Erickson et al., 2015; Nichols et al., 2015). More than four decades since the passing of the Clean Air Act Amendments of 1970, it has been demonstrated that the economic welfare and growth can improve while emissions are decreasing. For countries that have more serious air quality issues, the public health and welfare is even more heavily affected. Estimated deaths from lung cancer were 17, 200, and 689 per 10,000 for lifetime occupational exposures of 1, 10, and 25 μg/m^3, respectively, for elemental carbon such as in diesel exhaust (Vermeulen et al., 2014).

The transition to EVs and SPCSs has the potential to improve air quality and reduce the number of early deaths due to air pollution from mobile sources. Caiazzo et al. (2013) have reported that the estimated early deaths annually from road transportation are about 53,000 due to particulate matter and about 5300 due to ozone in the United States. In addition, they show about 52,000 early deaths from particulate matter and 1700 from ozone that are attributed to oil and coal-based electric power generation in the United States.

Globally, Lelieveld et al. (2015) estimate 3.3 million air quality related deaths annually (mostly due to particulate matter and ozone). Globally, about 5% of these are attributed to land transportation (165,000). The 5% value is small compared to the values reported by Caiazzo et al. (2013) and the OECD (2014) and WHO (2015b). Electric power generation is the third most significant source of air pollution after agriculture and natural sources in Lelieveld et al. (2015).

In a larger sense, all vehicles cause emissions of criteria pollutants and greenhouse gases. Well-to-wheel emissions include all air pollutant emissions associated with fuel production, processing, distribution, and use. Table 8.3 compares well-to-wheel greenhouse gas emissions for conventional gas and electric vehicles for a 100-mile trip for the United States, based on average electricity sources (i.e., coal, gas, nuclear, etc.). Emissions for conventional gas vehicles include those from extraction, refining, distribution, and use of fuel. Although all-electric vehicles do not produce tailpipe emissions, the electricity that is used to charge the battery might produce air emissions in being generated. If the electricity used to charge an all-electric vehicle

TABLE 8.3
Well-to-Wheel Emissions of Greenhouse Gases for a 100-Mile Trip

Vehicle (Compact Sedans)	Estimated Greenhouse Gas Emissions ($lb\ CO_2$ Equivalent)
All-electric vehicle	54
Plug-in hybrid electric vehicle	61
Hybrid electric vehicle	51
Conventional gas vehicle	99

Source: U.S. Department of Energy. 2015. Emissions from hybrid and plug-in electric vehicles. Alternative Fuels Data Center. http://www.afdc.energy.gov.

comes from a nonpolluting, renewable source, such as wind or solar, driving the vehicle produces no emissions.

There is a need for the continuing reduction of emissions associated with transportation and the generation of electricity. There is evidence to suggest that such action will have economic benefits. In the United States, the potential value of the air quality improvements for human health alone totals $37–90 billion each year (USEPA, 2015e). The total health costs associated with air pollution in China have been estimated to be about 1.2–3.3% of China's gross domestic product (Kan et al., 2009). Coal-fired power plants and transportation are major sources of China's air pollutants. Thus, SPCSs and EVs have the potential to dramatically improve air quality in China. There is progress with respect to the installation of SPCSs in China (Ho, 2015). In terms of alternative energy and specifically the use of SPCS, solar energy prices have decreased while gasoline prices have been increasing during the last 50 years. An estimated $188 million was spent as a result of pollution-related healthcare in the state of California alone, just between 2005 and 2007 across Medicare, Med-Cal, and private insurers (Romley et al., 2010). In addition, millions of lost work and school days are the result of unacceptable pollution levels.

The Organization for Economic Cooperation and Development (OECD, 2014) estimates that air pollution due to road transportation costs $850 billion per year for OECD countries. A WHO report (WHO, 2015b) estimates air pollution costs at $1.6 trillion per year for European countries, and that about 50% of these costs are attributed to road transportation. China and India are not included in either of these estimates; the OECD report estimates $1.7 trillion per year for China and $500 billion per year for India for all air pollution costs. The percentage of those costs attributed to road transportation is substantial but may be less than 50% for these countries (OECD, 2014).

Often neglected are the more distant external costs, such as the price of the military in securing foreign oil, the cost of catalytic converters, and the removal of sulfur from diesel. These are also real costs, however, and moving away from traditional ICE vehicles reduces those costs.

When powered by electricity from natural gas or WWS (renewables such as wind, water, and solar) rather than coal, electric vehicles can reduce NO_x,

PM_{10}, VOCs, and CO in urban areas to meet WHO standards (Tessum et al., 2014; Nichols et al., 2015). Unlike conventional vehicles, the majority of life-cycle emissions of EVs are in manufacturing, far away from cities, instead of at the point of usage (i.e., on a roadway); this is better for urban environments (Tessum et al., 2014; Nichols et al., 2015). The environmental impacts of EV manufacturing can similarly be reduced by using electricity generated with wind and/or solar energy (Tessum et al., 2014).

The air quality in urban environments impacts the quality of life for people who live in or near the center of large cities. There is significant value associated with the reduction or elimination of air pollutants associated with mobile sources and coal-fired power plants. The installation of large numbers of SPCSs and the electrification of transportation has the potential to impact the quality of life by improving air quality. Soret et al. (2014) have reported on estimated improvements in air quality from the electrification of transportation in Barcelona and Madrid, Spain.

Currently, in the United States, there is a federal tax rebate of up to $7500 on EVs and PHEVs, and also decreased cost of installing a Level 2 charging station, along with many other state-specific incentives. Direct benefits include savings in energy costs: while a conventional light duty vehicle's energy costs amounts to $1955/year, it is only $1004 for hybrid electric vehicles, and $370/year for battery electric vehicles (Erickson et al., 2015). With promising battery technology on the horizon, the price of batteries is going down, adding to the viability of widespread EV use (Nykvist and Nilsson, 2015).

California in particular is making an effort to improve air quality by encouraging EVs and solar energy. There are incentives associated with EV purchases and use as well as regulatory controls. California clean vehicle rebates are $1500 for the Chevy Volt and many plug-in hybrids, and $2500 for the Ford Focus and other EVs that are powered electrically (CVRP, 2015; Drive Clean, 2015). In 2014, about 10% of new car sales in California were EVs (EV News, 2014). California also has a zero-emissions mandate for auto dealers: to sell an increasing percentage as EVs that have zero emissions when operated with electrical power. This mandate will require about 270,000 EVs for the 2025 model year (CEPA, 2015; O'Dell, 2015). The mandate begins in 2018 and the percentage increases each year until 2025 (CEPA, 2015).

There are many incentives related to EVs and SPCSs in California and other locations. Many of these incentives are associated with efforts to improve air quality. There are state incentives to generate electricity with solar panels in California and some other states (CPUC, 2015; DSIRE, 2015). In addition, there are incentives to install electric vehicle supply equipment to charge EVs (AFDC, 2015; Berman, 2015). It is important to have an infrastructure that allows EV owners to charge their vehicles conveniently. In Colorado, there are state incentives to purchase EVs, which do not need to be tested for emissions. Air quality in the Denver area can be improved by replacing vehicles that use gas or diesel with EVs that are charged at SPCSs. There is a low emission vehicle sales tax exemption in Colorado (Colorado, 2015), and there are SPCSs in several cities in Colorado. Minnesota has a zero emissions charging challenge (ZECC) program to encourage the use of wind and solar energy for EV charging (Drive Electric Minnesota, 2015). There are SPCSs at a number of locations in Minnesota; most are in the metro area in St. Paul and Minneapolis (Drive Electric Minnesota, 2015). The American Lung Association in Minnesota is one of the partner organizations of Drive Electric Minnesota.

The electric power from SPCSs is much cleaner than that from coal-fired power plants. The benefits of reducing pollution associated with coal are substantial (USEPA, 2011). Various countries, including Japan, France, Norway, the Netherlands, and China, have developed incentive programs similar to those in the United States and in individual U.S. states (NRC, 2015). Incentives include tax exemptions or rebates, exemptions or reductions in registration fees, and reduced roadway taxes or tolls.

8.5 Conclusions

Air pollution has continued to pose risks to human health and well being, resulting in the promulgation of increasingly stringent regulations. The benefits of implementing air quality regulations have greatly outweighed their costs (USEPA, 2015a). These regulations have resulted in major emissions reductions and improvement in ambient air quality. However, more sustainable solutions, including electrification of transportation, are needed. Air quality in urban areas will continue to improve with the adoption of EVs because they have significantly lower emissions than conventional vehicles, especially when powered by electricity generated by solar energy. This improvement will also reduce mortality and health problems due to air pollution, which continues to afflict urban environments. Greater adoption of EVs will help urban areas meet NAAQS for CO, NO_2, ozone, and $PM_{2.5}$ and decrease greenhouse gas emissions. The quality of urban life would be dramatically better in many cities if all transport used EVs powered with wind and solar energy.

References

AFDC. 2015. Alternative Fuels Data Center. U.S. Department of Energy; http://www.afde.energy.gov.

American Lung Association. 2015. State of the Air Report; http://www.stateoftheair.org/.

Apte, J.S., J.D. Marshall, A.J. Cohen, and M. Brauer. 2015. Addressing global mortality from ambient PM2.5. *Environ. Sci. Technol.* 49: 8057–8066.

Berman, B. 2015. Incentives for plug-in hybrids and electric cars, plug-in cars; http://www.plugincars.com/federal-and-local-incentives-plug-hybrids-and-electric-cars.html.

Caiazzo, F., A. Ashok, I.A. Waitz, S.H.L. Yim, and S.R.H. Barrett. 2013. Air pollution and early deaths in the United States, Part I: Quantifying the impact of major sectors in 2005. *Atmospheric Environment* 79: 198–208.

California. 2014. California vehicle grid integration (VGI) roadmap: Enabling vehicle based grid services; http://www.caiso.com/.

CEPA. 2015. Zero-emission vehicle legal and regulatory activities and background, CEPA Air Resources Board; http://www.arb.ca.gov/msprog/zevregg/zevregs.htm.

Colorado. 2015. Green driver state incentives in Colorado; http://www.dmv.org/co-colorado/green-driver-state-incentives.php.

CPUC. 2015. About the California solar initiative, California Public Utilities Commission; http://www.cpuc.ca.gov/puc/energy/solar/aboutsolar.htm.

CVRP. 2015. CVRP Final Report 2013-2014, Clean Vehicle Rebate Project, California Air Resources Board; https://cleanvehiclerebate.org/eng/content/cvrp-final-report-2013-2014.

Drive Clean. 2015. Drive Clean, Plug-In Electric Vehicle Resource Center, 2015; http://driveclean.ca.gov/pev/costs/vehicles.php.

Drive Electric Minnesota. 2015. Drive Electric Minnesota; http://www.driveelectricmn.org.

DSIRE. 2015. Database of state incentives for renewables and efficiency, N.C. Clean Energy Technology Center, North Carolina State University; http://www.dsireusa.org.

Erickson, L.E., A. Burkey, K.G. Morrissey et al., 2015. Social, economic, technological, and environmental impacts of the development and implementation of solar-powered charge stations. *Environmental Progress & Sustainable Energy*. doi 10.1002/ep.

EV News. 2014. Electric vehicle news for November 27, 2014; http://www.electric–vehiclenews.com/.

Ho, V. 2015. China starts building its largest electric car solar charging complex. *Mashable*; http://mashable.com/2015/10/21/China-electric-car/.

Kan, H., B. Chen, and C. Hong. 2009. Health impact of outdoor air pollution in China: Current knowledge and future research needs. *Environmental Health Perspectives* 117: A187.

Lelieveld, J., J.S. Evans, M. Fnais, D. Giannadaki, and A. Pozzar. 2015. The contribution of outdoor air pollution sources to premature mortality on a global scale, *Nature* 525: 367–371.

Lim, S., J. Vos, A.D. Flaxmon et al. 2012. A comparative risk assessment of burden of disease and injury attributable to 67 risk factors and risk factor clusters in 21 regions, 1990–2010: A systematic analysis for the Global Burden of Disease Study 2010. *The Lancet*, 380: 2224–2260.

Lurmann, F., E. Avol, and F. Gilliland, 2015. Emissions reduction policies and recent trends in Southern California's ambient air quality. *Journal of the Air & Waste Management Association*, 65(3): 324–335.

NAAQS. 2012. http://www.plantservices.com/articles/2012/09-strategics prepare-naaqs-revisions/.

Nichols, B.G., K.M. Kockelman, and M. Reiter. 2015. Air quality impacts of electric vehicle adoption in Texas. *Transportation Research Part D: Transport and Environment*, 208–218.

Norway. 2012. Norway EVs and Clean Air European Association for Battery, Hybrid and Fuel Cell Electric Vehicles. AVERE. 2012-09-03. Norwegian Parliament extends electric car initiatives until 2018.

Norway, 2014. Air Pollution Fact Sheet, 2014, Norway. European Environment Agency; http://www.eea.europa.eu/themes/air/air-pollution-country-fact-sheets-2014.

NRC. 2015. *Overcoming Barriers to Deployment of Plug-In Electric Vehicles*. National Academies Press, Washington, DC.

Nykvist, B. and M. Nilsson, 2015. Rapidly falling costs of battery pack for electric vehicles, *Nature Climate Change* 5: 329–332.

O'Dell, J. 2015. Will California's zero-emission mandate alter the car landscape? *Edmonds*, May 27, 2015; http://www.edmonds.com/fue/-economy/will-californias-zero-emissions-mandate-alter-the-car-landscape?/.

OECD. 2014. The cost of air pollution: Health impacts of road transport, organization for economic cooperation and development report, OECD Publishing, doi: 10.1787/9789264210448-en; http://www.oecd.org.

Romley, J.A., A. Hackbarth, and D.P. Goldman. 2010. The impact of air quality on hospital spending. RAND Corporation, Santa Monica, CA; http://www.rand.org/pubs/technical_reports/TR777.

Soret, A., M. Guevara, and J.M. Baldasano. 2014. The potential impacts of electric vehicles on air quality in the urban areas of Barcelona and Madrid (Spain), *Atmospheric Environment*, 99: 51–63.

Tessum, C.W., J.D. Hill, and J.D. Marshalla. 2014. Life cycle air quality impacts of conventional and alternative light-duty transportation in the United States. *PNAS*. 111: 18490–18495.

U.S. Department of Energy. 2015. Emissions from hybrid and plug-in electric vehicles. Alternative Fuels Data Center. http://www.afdc.energy.gov.

USEPA. 2008. Plain English guide to the Clean Air Act: Cars, Trucks, buses, and "nonroad" equipment. Aug. 29, 2008. http://www.epa.gov/air/caa/peg/carstrucks.html.

USEPA. 2011. Benefits and Costs of Cleaning Up Toxic Air Pollution from Power Plants. EPA Fact Sheet; http://www3.epa.gov/mats/pdfs/20111221MATSimpactsfs.pdf.

USEPA. 2015a. Progress Cleaning the Air and Improving People's Health. EPA. http://www.epa.gov/air/caa/progress.html.

USEPA. 2015b. Air Quality Trends. EPA. http://www.epa.gov/airtrends/aqtrends.html#comparison.

USEPA. 2015c. National Ambient Air Quality Standards (NAAQS); http://www3.epa.gov/ttn/naaqs/criteria.html.

USEPA. 2015d. National Ambient Air Quality Standards for Ozone; http://www3.epa.gov/airquality/ozonepollution/actions.html.

USEPA. 2015e. Air Emission Sources; http://www3.epa.gov/air/emissions/index.htm.

Vermeulen, R., D. Silverman, E. Garshick, J. Vlaanderen, L. Portengen, and K. Steenland, 2014. Exposure-response estimates for diesel exhaust and lung cancer mortality based on data from three occupational cohorts. *Environ. Health Perspect*. 122(2): 172–177.

Wadud, Z. and I.A. Waltz. 2011. Comparison of air quality-related mortality impacts of different transportation modes in the United States. *Transportation Research Record*, 2233: 99–109.

WHO. 2008. World Health Organization estimates in EEA-32 obtained from WHO Global Health Observatory Database http://app.who.int/ghodata/?vid=34300.

WHO. 2014. World Health Organization. http://www.who.int/mediacentre/news/releases/2014/air-pollution/en/.

WHO. 2015a. Ambient (Outdoor) Air Quality and Health, Fact Sheet No. 313; http://www.who.int/mediacentre/factsheets/fs313/en/.

WHO. 2015b. Economic Cost of the Health Impact of Air Pollution in Europe. World Health Organization Report; www.euro.who.int/.

9

Economics, Finance, and Policy

**Blake Ronnebaum, Larry E. Erickson, Anil Pahwa,
Gary Brase, and Michael Babcock**

CONTENTS

Education is the best economic policy that there is.

Tony Blair

9.1 Introduction

The electrification of transportation in the United States is a quickly growing industry that has several implications for the country's economy as well as its citizens. The emergence and adoption of hybrid-electric vehicles (HEVs), plug-in hybrid-electric vehicles (PHEVs), and battery-electric vehicles (BEVs) will significantly change the country's energy infrastructure and economy. In 2013, an average of 8.774 million barrels of finished motor gasoline was used each day (U.S. Energy Information Administration, 2014f). With an increasing population of electrically powered vehicles on the road, accompanied by a gradual increase in fuel economy, the amount of gasoline and oil consumed in the United States will decrease. The rate at which that decrease occurs greatly depends on the rate of adoption of electric vehicles. Estimates vary as to how many electric vehicles will be on the road by 2030. Some optimistic experts posit that a 90 percent market share will occur by 2030 (including HEVs, PHEVs, and BEVs) (up to 90% of US cars could be "green" vehicles by 2030, 2011), whereas others estimate that the market share could be as low as 6 percent (U.S. Energy Information Administration, 2014c). This chapter reviews the costs, benefits, and economics of conventional vehicles versus electric vehicles, and how those economics are affected by both conventional power generation to charge electric vehicles versus the development of solar powered infrastructures to charge electric vehicles.

The large-scale addition of electric vehicles to the transportation system will impact the demands and characteristics of the power grid infrastructure, and many of these impacts are potentially problematic. Charging electric vehicles using electricity generated through solar panels would be much more cost-effective than charging directly from the grid, not to mention better for the environment. The solar panels would also help reduce stress placed on the power grid during peak hours, such as hot summer days when electricity costs can increase to several hundred dollars or sometimes $1000 per megawatt hour (ISO New England Inc., 2003). With more charging availability for EV owners, it will be easier to avoid charging at times of peak power demand. Therefore, it is important to understand the internal and external economic impacts of using electric vehicles versus conventional vehicles, and it is also important to understand how those economics are altered by including EVs charged through solar powered charging stations (versus grid-supplied charge stations in the present market). Research has previously been done to determine the economic feasibility of installing SPCSs in parking structures (Goli and Shireen, 2014; Jamil et al., 2012; Tulpule et al., 2013; Zhang et al., 2013). This chapter seeks to focus on the indirect economic cost savings by switching to electric vehicles and solar power, as well as to provide an economic viability analysis for SPCSs, particularly using an example of a large research university in the American Midwest.

9.2 External Costs of Conventional Vehicles

Comprehensive and exact assessments of the external effects caused by consumption of fossil fuels are very difficult to measure and define. However, these effects are estimated to be very significant in the areas of human health, foreign affairs, and the environment.

9.2.1 Human Health

As reported in Chapter 8, road transport has been estimated to cause the highest number of pollution-related premature deaths—one report found that as many as 12,000 to 42,300 people died prematurely from road transport pollution in 2000 (Wadud, 2011). About 3900–12,000 of these deaths are specifically related to passenger car pollution, which is equal to about 10%–30% of the deaths caused by accidents in road transport in 2000. Passenger car-related pollution causes more damage compared to other types of transport because there is a higher density of passenger cars that are closer to people, as compared to other forms of transportation, such as aviation and shipping (Wadud, 2011). Diesel exhaust has also been classified as a carcinogen and gasoline exhaust as a possible carcinogen (Benbrahim-Tallaa et al., 2012). Measuring the cost of these deaths in monetary terms is very difficult. Small and Kazimi (1995) identify one method in which they evaluate peoples' willingness to pay to reduce their risk of death annually. In developed countries like the United States, people have reported to be willing to pay thousands of dollars each year to reduce this risk, resulting in a valuation by researchers of between $2.1 million and $11.3 million per life in 1992 dollars ($3.44 million and $18.49 million in 2012 dollars) (Small and Kazimi, 1995). Relating these estimates in value of life to the number of lives lost, the external costs of loss of life due to premature death from passenger car pollution yields estimates ranging from as little as $13.42 billion to as much as $221.9 billion per year. However, these estimates have probably decreased in recent years due to rising fuel costs, increased fuel economy in cars, and early adoption of EVs. In particular, the average fuel economy of passenger vehicles purchased increased from 28.5 to 35.2 miles per gallon from 2000 to 2012, meaning that emissions from passenger cars have likely decreased since 2000 (Davis et al., 2013). This indicates that the total number of deaths due to passenger car emissions would be fewer today than in 2000 because of reduced emissions. Nonetheless, these estimated costs are strikingly large.

The airborne pollutants that come from fossil fuel combustion are proven to have a wide range of effects on human health, and energy production and ground transportation are the main drivers of these emissions. For example, ozone is a dangerous pollutant created from reactions of organic compounds and nitrogen oxides produced from fossil fuel combustion. Inhalation of small amounts of ozone can cause respiratory health issues (Smith et al., 2009). In terms of air pollution caused by transportation, it is estimated that air pollution costs about 1.1 cents per vehicle mile traveled in 1990 (Victoria Transport Policy Institute, 2009), which would probably be about the same amount today after accounting for both inflation as well as increased fuel economy. This cost encompasses the number of people exposed to the pollution, their mortality and morbidity due to exposure, and the value placed on human life. With 232 million vehicles on the road in the United States in 2012, this amounts to nearly $79.1 million per day or $28.9 billion per year (Davis et al., 2013). This value falls within the range of the estimated cost of premature deaths reported earlier. Water pollution is also an adverse effect of vehicle use, resulting from de-icing roads, oils and other fluids leaking into water sources, and the settlement of air pollution into water sources. The cost of water pollution from average U.S. vehicles is estimated to be around 1.4 cents per vehicle mile or about $18.4 billion per year (Victoria Transport Policy Institute, 2009).

9.2.2 Foreign Affairs

In 2012, the United States' net imported oil equaled about 7.4 million barrels per day. This equaled about 40% of the oil used in 2012 (U.S. Energy Information Administration, 2013). Most of this oil is imported from the western hemisphere, where no significant conflict takes place in order to obtain it securely. Twenty-eight percent of net petroleum imports come from the Persian Gulf, where military conflict is a significant concern. It's estimated that military costs are around $500 billion per year, or about $140 per imported barrel (Victoria Transport Policy Institute, 2009). About 45% of oil was refined to produce motor gasoline in 2013 (U.S. Energy Information Administration, 2014b), which accounts for $225 billion of the military costs for the Persian Gulf, if we assume that 45% of the oil imported from that region is also refined into motor gasoline. In order to combat high oil imports, the government has employed the Corporate Average Fuel Economy (CAFE) program. This program sets a standard that requires all cars sold to have a minimum average fuel economy. This number has slowly increased from 18 miles per gallon when it was first adopted in 1975 to 35 miles per gallon in 2016 in order to reduce our dependence on oil (Horn and Docksai, 2010). By reducing gasoline consumption by 10–30%, the costs of military presence in the Persian Gulf to secure our oil demand could be reduced by $22.5–67.5 billion dollars per year. These estimates, however, do not include costs (or cost savings) that could be incurred due to soldier injuries, such as disability costs. Additionally, these effects would also reduce emissions of greenhouse gases from oil tankers and from reduced operation of military vehicles.

9.2.3 Environment

The environment is of great concern when it comes to consuming fossil fuels. Many of the byproducts of fossil fuel combustion have damaging effects on the environment, the most notable being climate change. Furthermore, it's not just the end-use combustion of fuel that can damage the environment. The Deep Water Horizon oil spill that occurred in 2010 caused massive amounts of ecological damage in the Gulf of Mexico. The costs to clean up the spill, along with compensations, are estimated to be about $20–40 billion dollars (Victoria Transport Policy Institute, 2009). However, this only includes costs that could be determined on paper. Environmental and aesthetic costs, such as damaged landscapes and endangered wildlife, are difficult to estimate.

These costs are roughly estimated to be between $10 and $30 billion each year (Victoria Transport Policy Institute, 2009).

The transportation and energy sectors account for the most emissions of greenhouse gases in the United States, making them the biggest contributors to climate change. In June 2014, President Obama announced a plan to reduce carbon dioxide emissions across the United States by 30% relative to 2005 levels (Plumer, 2014). In order to see this plan to fruition, the energy and transportation sectors are the most important areas to begin reducing emissions. One way to place a value on the price of greenhouse gas emissions is to look at the cost for countries to purchase carbon dioxide certified emissions reductions (CERs), which countries can use to meet their emissions targets and to combat climate change. In 2012, these CERs cost about $5.90 per metric tonne of carbon dioxide emissions ($0.0059 per kilogram). By comparing this price to the amount of carbon dioxide emissions released by conventional vehicles in 2012, a cost of climate change can be placed on emissions.

9.2.4 Total External Costs of Conventional Vehicle Emissions

We combine all of these costs to find the external costs of emissions per light duty vehicle per year as well as the cost per mile traveled for individual vehicles. We also assume that all 232 million light-duty vehicles on the road in the United States release the same amount of emissions every year and all travel the national average of 11,300 miles per year (Davis et al., 2013). The costs due to climate change are $28.17 per vehicle/year, or 0.25 cents per mile. Costs due to environmental destruction (aesthetic damages and species loss) are estimated to be $129–301 per vehicle per year or about 1.1–2.7 cents per vehicle mile traveled. Combined air and water pollution costs based on the aforementioned values are estimated to be around $282.50 per vehicle/year, or 2.5 cents per vehicle mile traveled. Costs of military security of foreign oil are estimated to be around $2150 per vehicle/year, or about 19 cents per vehicle mile traveled. In total, these external costs come to be about $2670 per vehicle per year, or about 24 cents per vehicle mile traveled. Of course, these values depend on vehicle specifications, location, and several other variables, but they represent a good approximation of the economic costs associated with emissions released from conventional vehicles. To what extent could these costs be mitigated by the use of electric vehicles? Furthermore, how would the addition of electric vehicles charged using SPCSs further alter these economics?

9.3 External Costs of Electric Vehicles

Understandably, the external effects of driving electric vehicles as opposed to driving conventional vehicles are far less costly. Many of the external costs that come with gas powered vehicles are completely avoided, such as military costs to secure foreign oil. With electric vehicles charging from the grid, the only source of emissions would be from power plants and high emissions concentrations would be localized to a few areas, rather than widely dispersed via vehicles. However, some external costs still exist.

Sulfur dioxide, which comprises only a small fraction of greenhouse gases, not only can cause and exacerbate respiratory diseases such as asthma, but also significantly contributes to the formation of acid rain (Smith et al., 2009; United States Environmental Protection Agency, 2014). In 2012, electricity production generated 63.3% of sulfur dioxide emissions in the United States, mostly from coal combustion. However, the amount of sulfur dioxide emissions in the United States is quickly decreasing with increasing legislation and regulation (United States Environmental Protection Agency, 2014).

In 2012, the transportation sector released a total of 1837 teragrams (Tg) of CO_2e, which accounted for 28.2% of all greenhouse gas emissions that year. However, only 793.8 Tg CO_2e of that total were emitted from passenger cars (only 43.2%) (United States Environmental Protection Agency, 2014). Table 9.1 shows the reduction of greenhouse gases due to increased use of PHEVs and BEVs in the United States, assuming the baseline decrease in emissions observed from 2008–2012 of 2.95% remains constant, likely due to the increase in fuel economy for passenger vehicles (United States Environmental Protection Agency, 2014). The data results from six scenarios: a baseline, where the current amount of electric vehicles stays constant, and 10%, 20%, 30%, 40%, and 50% market shares of EVs by 2030. Plug-in hybrids have been given a weight of 0.6 in this table, assuming that they drive using only electricity an average of about 60% of the time (Goldin et al., 2014). The growth of PHEVs is shown to be larger than that of BEVs because PHEVs cost less and do not have the range of anxiety issues that BEVs have, although those issues are expected to be less severe for BEVs by 2030 due to anticipated improvements in battery technology and charging infrastructure. For the most optimistic scenario of a 50% market penetration of EVs by 2030, the amount of greenhouse gases emitted could be reduced by about 42% from passenger vehicles compared to 2012. This optimistic measurement would coincide with an 18.1% decrease in greenhouse gas emissions across the entire transportation sector, assuming all else remains equal. However, this data does not account for the increased energy generation needed to provide power to the increasing numbers of electric vehicles. If all the energy for EVs comes from SPCSs, the carbon emissions would be small. Without a growth in much less carbon-intensive energy production, the increase in EVs will do less to reduce carbon emissions. The table also shows the cost of carbon emissions determined by the cost of certified emissions reductions ($0.0059 per kg). The data shows that, with a 30% EV market share in 2030, the cost of the amount of carbon emissions released can be reduced by over $1 billion in 2012 dollars compared to the baseline estimate.

TABLE 9.1

Reduction of Greenhouse Gas Emissions in Regards to Adoption of Electric Vehicles and Decreased Gasoline Use

	BEV Market Share in 2030	PHEV Market Share in 2030	Total Percent Decrease in Emissions Every Four Years	Total GHG Emissions from Passenger Cars in 2030 (Tg CO_2e)	Percent Reduction in Passenger Car Emissions (CO_2e) from 2012	Cost of Emissions in 2030[a] (Millions of Dollars)
Baseline	0%	0%	2.95%	688.4	13.28%	$4061
10% EV market share	3%	7%	4.55%	631.2	20.48%	$3724
20% EV market share	6%	14%	6.15%	574.1	27.68%	$3387
30% EV market share	10%	20%	7.85%	513.7	35.28%	$3031
40% EV market share	15%	25%	9.61%	450.5	43.25%	$2658
50% EV market share	20%	30%	11.39%	386.9	51.26%	$2283

[a] Values are in 2012 dollars.

By simply switching from conventional vehicles to EVs, the number of greenhouse gas emissions produced each year can decrease significantly. Table 9.1 shows how increases in the market shares of EVs by 2030 could decrease the amount of greenhouse gas emissions in that time. With the current baseline trend, emissions from passenger vehicles can be expected to decrease by 12.6% from 2012 to 2030. When we include a 20% market share of EVs, emissions are expected to be reduced by nearly twice that amount. However, this data only takes into account emissions from the vehicles directly. The data does not take into account the amount of greenhouse gases released from generating electricity to power the new share of electric vehicles.

The great challenge for electric vehicles having an indisputable positive impact on the environment, however, is that if the energy produced to power EVs remains as fossil fuels (e.g., coal), there is a smaller environmental benefit to driving one. Therefore, a consumer could potentially be better off purchasing a cheaper conventional vehicle. In order for EVs to have a more positive effect on the environment, the green-energy infrastructure and EV charging infrastructure of SPCSs must grow at the same rate to accommodate the demands of consumers.

In 2012, the U.S. energy generation mix consisted of 68.5% fossil-fuel energy, and the remaining 31.5% was produced mostly using nuclear and renewable sources (U.S. Energy Information Administration, 2014a). If EVs are charged from the grid, the emissions released from generating that electricity is about 0.500 kg of CO_2e per kWh, which would be about 5.17 kg of CO_2e per day of driving (U.S. Energy Information Administration, 2014e; United States Environmental Protection Agency, 2014). Although these are greatly reduced emissions compared to the emissions associated with a gallon of gasoline (about 8.788 kg of CO_2e per gallon, 13.1 kg of CO_2e per day per vehicle (Davis et al., 2013)), it is still not a zero-emission vehicle. Therefore, since the emissions per day of electric vehicles are only 39.4% of those released by conventional vehicles, we can assume that the external costs of air pollution and climate change for EVs charged from the grid are only 39.4% of the costs for conventional vehicles. This means that air pollution costs for EVs charged from the grid are 0.43 cents per mile and climate change costs are only 0.099 cents per mile, to give a total of 0.529 cents per mile for the cost of indirect emissions from EVs.

Similarly, water pollution is still a problem for electric vehicles, although it is still less severe than it is for conventional vehicles. This pollution comes mostly from de-icing roads in winter months. The Victorian Transport Policy Institute estimates that the average cost of water pollution in 2012 dollars would be about 0.775 cents per mile or almost $90 per year (Victoria Transport Policy Institute, 2009).

Because there is no military conflict involved in securing fuel for electric vehicles and no direct environmental destruction taking place to power those vehicles, the environmental and foreign affairs external costs are essentially

TABLE 9.2

Reduction of Gasoline Consumption in Relation to the Adoption of Electric Vehicles by 2030

	PHEV Market Share in 2030	BEV Market Share in 2030	Decrease in Gasoline Consumption Every Five Years	Gasoline Consumption (Million Barrels/Day) in 2030	Gasoline Saved in 2030 Compared to 2012 (Million Barrels/Day)
Baseline	0%	0%	2%	8.079	0.695
10% EV market share	7%	3%	5%	7.500	1.274
20% EV market share	14%	6%	7%	6.951	1.823
30% EV market share	20%	10%	9%	6.407	2.367
40% EV market share	25%	15%	11%	5.867	2.907
50% EV market share	30%	20%	14%	5.179	3.595

zero. The amount of gasoline saved from switching to electric vehicles is outlined in Table 9.2. PHEVs have been given a weight of 0.6 since they use electricity about 60% of the time. We also include a baseline of a 2.4% decrease in gasoline consumption every five years, which is the decrease in consumption that was observed from 2008 to 2013 (U.S. Energy Information Administration, 2014f). This decrease is most likely due to increased vehicle fuel economy. The results show that even with a small increase of having a 10% market share of electric vehicles in 2030, the amount of gasoline consumed per day decreases by over one million barrels per day, which would reduce our dependence on foreign oil for gasoline from 8.774 million barrels per day in 2013 to 7.5 million barrels per day in 2030.

After adding all of these factors, the total external costs of an electric vehicle are 1.30 cents per mile for a vehicle charging from the grid, or about $146.90 per year—only 7% of the external costs of a conventional vehicle.

9.4 Economics of EV Adoption

Although the economic benefits of purchasing EVs are outstanding in the long run, BEVs come with their share of downsides in the current market. High initial purchasing prices and more restricted operating range represent the largest contributors to consumers' hesitance to purchase BEVs (Bullis, 2013). However, the cost of EVs is going down and EV sales are increasing quite rapidly. According to the Electric Drive Transportation Association, the number of year-to-date plug-in vehicle sales increased by 35% from May 2013 to May 2014, and the number of total plug-in vehicles on the road increased by about 102% in the same period (Electric Drive Transportation Association, 2014). If the growth in sales continued annually, there would be approximately 63 million electric vehicles sold by 2030—almost 30% of all vehicles on the road in 2012. Recently, some cities have been able to encourage large-scale EV adoption through several incentives.

Atlanta, Georgia has become the second-largest market in the United States for battery-electric vehicles (Los Angeles, California is the largest) by ameliorating these concerns about initial costs and operating range. In order to drive the total cost of the vehicles down, Atlanta's main power utility offers off-peak charging at a rate of 1.3 cents per kilowatt hour. When all these factors are considered, the cost to lease a Nissan Leaf can be as low as $28 per month. Atlanta also grants EV owners the permission to drive in high-occupancy vehicle lanes in order to avoid high traffic congestion, reducing range anxiety as the vehicle spends less time consuming energy stored in its battery. The effect of all of these incentives is significant: in 2014, the share of BEVs in Atlanta was 2.15% of registrations; over five times greater than the national average share. Coca Cola's headquarters in Atlanta has installed 75 EV charging stations for its estimated 100 employees who own electric vehicles, further incentivizing the adoption of EVs in the city (Ramsey, 2014). In addition, to further combat high initial prices for the vehicles, the state of Georgia offers a tax credit greater than $4000 for BEV purchases. More issues of policy regarding SPCSs will be discussed later in the chapter.

The rate of technology growth in the electric vehicle market is also something to consider when forecasting the adoption rate of EVs, especially BEVs. A main source of worry in BEVs is the capacity of the battery. The most economical BEV right now is the Nissan Leaf, which can travel 84 miles on a single charge; however, vehicles with more expensive and larger batteries, such as those found in Tesla Motors' Model S (60 kWh battery pack), which boasts a range of 208 miles, also costs over twice as much as the Nissan Leaf. Even though the Model S boasts a higher capacity and range, the Leaf is slightly more efficient in terms of fuel costs when compared to the average new conventional vehicle (U.S. Department of Energy, 2014a). The Model S's increased battery capacity is a result of a large improvement in battery technology. This technology continues to be improved as scientists and engineers develop new materials for batteries; recent prices of $300 per kilowatt hour of storage have been reported (Nykvist and Nilsson, 2015). For example, a combination of a lithium-air battery and an aluminum-air battery designed by the energy company Phinergy powered an EV for 1800 miles (3000 km) on a single charge. This battery also weighs five times less than the battery used in the Model S, making the vehicle significantly more efficient (Rosen, 2014). Additionally, Tesla CEO Elon Musk released the patents for their "Supercharger" charging stations and other technologies to the public in June 2014, encouraging collaboration and development from companies across the market (Smith, 2014). This will streamline the production of charging stations for electric vehicles, as well as decrease their costs as more companies begin to employ these technologies. This development is also notable since many of the Superchargers are powered by solar panels, providing clean power and shade to any vehicles being charged, which has intrinsic value. As these technologies become better understood and improved, they will become cheaper so that average consumers can utilize them.

There are direct economic benefits to consumers for switching from gas to electric fuel. In 2012, 232 million light-duty vehicles were in use in the United States with an average fuel economy of 20.8 miles per gallon. Each vehicle traveled an estimated 11,300 miles per year, or about 31 miles per day (Davis et al., 2013). The average price of gasoline to power these vehicles in 2012 was $3.60 per gallon (Hargreaves, 2012). These costs come to $1955.77 per vehicle/year, or about $5.36 per day. Electric fuel economy for BEVs and PHEVs is around 3 miles per kWh (U.S. Department of Energy, 2014a), and the average price for electricity in the United States in 2012 was 9.84 cents/kWh (U.S. Energy Information Administration, 2014d). Using the same driving statistics as before, it is calculated that driving a BEV in 2012 would have cost $370.64 per vehicle/year, or about $1.02 per day. Assuming PHEVs run on electricity an average of 60% of the time they're driven, their cost is $1004.69 per vehicle/year, or $2.75 per day. As the initial price of purchasing EVs continues to decrease, they will become highly competitive in relation to costs in the long-run when compared to conventional vehicles. There's also the matter of maintenance. The cost to own a vehicle in 2012 after deducting fuel was close to $7000 (Hunter, 2012). For instance, diesel delivery trucks owned by Staples were traded in for electric trucks, which are predicted to cost almost 90% less in maintenance each year (Ramsey, 2010). The reduced maintenance costs of electric vehicles mostly stem from the fact that the electric motor has only one moving part and does not need oil or transmission fluid, making the yearly cost to own electric vehicles significantly less than the cost to own conventional vehicles.

9.5 Solar Powered Charge Stations (SPCSs)

The electrification of transportation is crucial to combat rising greenhouse gas emissions and reduce dependence on fossil fuels. As electric vehicles become more popular, however, we can expect an increased demand for electric power and this raises concerns about the consequences of pollution generated by power plants. Presently, most of the electricity generated in the United States is produced by burning coal and natural gas. By increasing the number of electric vehicles, the amount of gasoline burned will decrease; however, the amount of coal and natural gas burned to generate the electricity needed to power said vehicles may increase. This is the main challenge with electric vehicle adoption: although consumers do reduce their emissions, the energy generation sector may increase theirs.

A way to escape this dilemma is to foster an accompanying increase in the use of solar and wind energy generation to provide the energy for electric vehicles. Charging from a renewable source, such as an SPCS, would negate these external costs. This is why developments such as SPCSs are so important and effective: not only does an SPCS provide electricity for the consumer, but it also reduces the stress put on electric companies to charge electric vehicles. Additionally, when a car isn't using the charging station to power up, the station continues to produce electricity that can be sent into the grid to further reduce combustion and the emissions produced by power companies.

In 2012, the energy sector released a total of 2064 Tg of carbon dioxide equivalent (CO_2e) emissions, which accounted for 31.6% of emissions in 2012. Of the total 2064 Tg CO_2e, nearly all of these emissions (2022.6 Tg CO_2e) were directly involved with energy generation (United States Environmental Protection Agency, 2014). In the case of SPCSs, we assume that each charge station has solar panels that cover 27 m^2 (average parking space area), are 15% efficient (the panels convert 15% of incident sunlight into energy), and receive 4 hours of peak sunlight each day, and therefore produce a total of 16 kWh of energy each day (Erickson et al., 2012). Table 9.3 shows the economics of the energy produced from SPCSs if they were to be adopted in large numbers, assuming an average energy price of 10.08 cents per kilowatt hour, the average price of electricity across all sectors in 2013 (U.S. Energy Information Administration, 2014d). The panels do decrease in efficiency over time (usually about 20% of their original output is lost after 25 years). However, this loss in efficiency is approximately offset by the expected rising costs of electricity (ExxonMobil, 2014).

One way to encourage the adoption of SPCSs would be for the government to provide subsidies for the renewable energy industry. In markets where there are external costs providing subsidies for the electric vehicles is an appropriate policy. According to the Energy Information Administration, 55.3% of all federal energy production subsidies and support are received by renewable energy production. However, 42% goes only to wind energy (U.S. Energy Information Administration, 2011). If more subsidies were approved for solar energy, the technology would be adopted more quickly because it would be more affordable to consumers, and more research and development could take place to make the technology more efficient and effective. This could also be done through state governments, especially southern states where solar irradiation is relatively high, more energy could be collected, and shade has significant value. Subsidies for public charging stations would also be important and beneficial to EV adoption. As plug-in vehicles become more popular and more affordable, an infrastructure parallel to that which is currently in place for gas powered vehicles will need to be available

TABLE 9.3

Economics of Installing Solar Powered Charge Stations, Assuming All Other Factors Remain Equal

Number of Solar Powered Charge Stations (Millions)	Cost to Install Each Station	Total Cost of Installing Stations (Billions)	Total Energy Produced Each Day (TWh)	Total Energy Produced Each Year (TWh)	Revenue Generated Each Day (Millions of Dollars)	Revenue Generated Each Year (Billions of Dollars)	Payback Time (Years)
100	$10,000	$1000	1.6	584.4	$161	$58.91	17.0
100	$15,000	$1500	1.6	584.4	$161	$58.91	25.5
100	$20,000	$2000	1.6	584.4	$161	$58.91	34.0
150	$10,000	$1500	2.4	876.6	$242	$88.36	17.0
150	$15,000	$2250	2.4	876.6	$242	$88.36	25.5
150	$20,000	$3000	2.4	876.6	$242	$88.36	34.0
200	$10,000	$2000	3.2	1169	$323	$117.82	17.0
200	$15,000	$3000	3.2	1169	$323	$117.82	25.5
200	$20,000	$4000	3.2	1169	$323	$117.82	34.0

Note: Yearly calculations are made using an average energy production for each charge station of 16 kWh/day, a price of 10.08 cents/kWh for electricity, and each year consists of 365.25 days.

for electric vehicles as well. Currently, there is one gas station for every 2500 people in the United States (Horn and Docksai, 2010). In order to match that infrastructure, governments could reward gas stations for providing charge stations in order to accommodate EVs.

By installing SPCSs in homes or businesses to provide energy to electric vehicles, a lot of money can be saved by not using electricity from the grid. Additionally, with a working business model, some charge stations would be able to make a profit within a few years of operation. Currently, there are a few drawbacks with solar energy and other renewables that make it less reliable than fossil fuels. Low energy density, intermittent generation, location constraints, and aesthetic effects are all hindering the adoption of solar electricity generation (Markandya and Wilkinson, 2007). However, the technology is expected to have a lot more potential than can currently be accessed, so research and development are highly important for photovoltaic technology. In order for SPCSs to be viable for use on a national scale, they need to produce enough energy to pay for themselves before the panels need to be replaced. The lifespan of solar panels depends on their output after their installation. The standard rule is that a panel should be providing at least 80% of its original output before replacing it. Most of the time, panels last about 25 years before reaching this threshold. However, some panels have lasted almost 40 years without needing replacement (Waco, 2011). The panels also provide shade for the vehicles while they are charging, which has important economic and social values as well.

The amount of revenue an SPCS would be able to generate over its lifetime varies depending on quite a few variables. Most notably, the economic sector the charge station is used in is a strong factor of viability. For our calculations, we are using our standard parameters of a panel structure that is 15% efficient, covers an area of 27 m^2, and gets four hours of peak sunlight per day to produce an average of 16 kWh of energy each day (Erickson et al., 2012). We use installation costs of $10,000, $15,000, and $20,000. We use this range to take into account variation in markets due to differences in geography, government incentives, and labor markets throughout the country. In the third quarter of 2015, the reported average price for nonresidential solar installations was $2.07/W. For one parking space with 4 kW of capacity, $2.07/W × 4000 W = $8028 for the installed solar panels (GTM, 2015). This estimated cost does not include the cost of the charge station for the EV.

In the United States, the four economic end-use sectors for energy are residential, commercial, transportation, and industrial. Table 9.4 details the net gain of each station by sector and the payback time for stations of the three installation costs per parking space. These results show that the residential sector is the most viable sector to install SPCSs, offering $7707 in net gain even after covering the $10,000 installation cost. In fact, the residential sector gives a positive net gain for the $15,000 installation cost scenario as well. For the transportation and commercial sectors, a positive net gain is only realized for installation costs of $10,000 and $15,000. The industrial sector showed the

TABLE 9.4

Economics of Solar Powered Charge Stations by Economic Sector and Installation Cost Using Energy Prices from 2013

Economic Sector	Initial Cost of Installation	Price of Electricity ($/kWh)	Revenue Generated Each Year from Electricity Production	Payback Time (Years)	Net Gain over 25 Years
Residential	$10,000	0.1212	$708.29	14.1	$7707.32
	$15,000	0.1212	$708.29	21.2	$2707.32
	$20,000	0.1212	$708.29	28.2	None
Commercial	$10,000	0.1029	$601.35	16.6	$5033.69
	$15,000	0.1029	$601.35	24.9	$33.69
	$20,000	0.1029	$601.35	33.3	None
Transportation	$10,000	0.1028	$600.76	16.6	$5019.08
	$15,000	0.1028	$600.76	25.0	$19.08
	$20,000	0.1028	$600.76	33.3	None
Industrial	$10,000	0.0682	$398.56	25.1	None
	$15,000	0.0682	$398.56	37.6	None
	$20,000	0.0682	$398.56	50.2	None

Note: Assumes solar panels cover 27 m^2, receive four hours of peak sunlight per day, and operate at 15% efficiency to produce 16 kWh each day.

least viable outcome, offering a positive net gain only when installation costs were less than $10,000. One can also estimate the base price of electricity that could be charged to pay for the station in 25 years. These prices are $0.0684 per kWh for a $10,000 installation, $0.1027 per kWh for a $15,000 installation, and $0.1369 per kWh for a $20,000 installation. This could be accomplished by local power companies offering to register an ID card to use at the station. The prices would then be added on to the customer's monthly bill.

In the case of a large Midwestern research university, for example, an SPCS would not be economically viable unless the installation price was very low because the university pays only about 7 cents per kWh for electricity. However, if a maintenance or parking fee were charged to those using the station, it may become economically viable. In this university's parking garage, there is currently a charge station for EVs that charges $3 for the first hour of charging and $1 for each additional hour (Davis, 2014). To calculate the impact this would make in increasing the viability of the charge stations, we assume several different scenarios. First, we assume scenarios in which every week, an average of one car, three cars, and seven cars use the charge station. We then assume that each car charges for an average of one hour, two hours, and four hours each day. Using this data, shown in Table 9.5, we find that these extra costs make quite a big difference in some scenarios. The best-case scenario, where seven cars charge for four hours each week, utilizes all of the energy generated by the SPCS. For a $10,000 installation

TABLE 9.5

SPCS Viability Including Cost of Charging

Initial Cost of Installation	Vehicles Charging Each Week	Hours Spent Charging	Income Each Year	Payback Time (Years)	Net Gain over 25 Years
$10,000	1	1	$565.08	17.7	$4127
$10,000	1	2	$617.08	16.2	$5427
$10,000	1	4	$721.08	13.9	$8027
$10,000	3	1	$877.08	11.4	$11,927
$10,000	3	2	$1033.08	9.7	$15,827
$10,000	3	4	$1345.08	7.4	$23,627
$10,000	7	1	$1501.08	6.7	$27,527
$10,000	7	2	$1865.08	5.4	$36,627
$10,000	7	4	$2593.08	3.9	$54,827
$15,000	1	1	$565.08	26.5	None
$15,000	1	2	$617.08	24.3	$427
$15,000	1	4	$721.08	20.8	$3027
$15,000	3	1	$877.08	17.1	$6927
$15,000	3	2	$1033.08	14.5	$10,827
$15,000	3	4	$1345.08	11.2	$18,627
$15,000	7	1	$1501.08	10.0	$22,527
$15,000	7	2	$1865.08	8.0	$31,627
$15,000	7	4	$2593.08	5.8	$49,827
$20,000	1	1	$565.08	35.4	None
$20,000	1	2	$617.08	32.4	None
$20,000	1	4	$721.08	27.7	None
$20,000	3	1	$877.08	22.8	$1927
$20,000	3	2	$1033.08	19.4	$5827
$20,000	3	4	$1345.08	14.9	$13,627
$20,000	7	1	$1501.08	13.3	$17,527
$20,000	7	2	$1865.08	10.7	$26,627
$20,000	7	4	$2593.08	7.7	$44,827

Note: First hour costs $3, and each additional hour costs $1. Value of electricity is 7 cents per kWh and solar panels produce 16 kWh of energy each day.

cost, this scenario lowers the payback time to a little less than four years with a net gain of almost $68,000 over 25 years. However, in the current market of electric vehicles, this best-case scenario is highly unlikely to occur. This method of payment, however, does improve the payback times of the charge station significantly—by about 10 years for each installation cost. However, the worst case scenario of one vehicle charging for one hour each week and an installation cost of $20,000 still does not garner a net gain in 25 years. At 7 cents per kWh, the income for one parking space from selling all of the electricity produced by the solar panels to the university would be $10,227 over 25 years. If there would be an added charge of $30 each year for a permit that included the opportunity to park in the shade of the solar panels, this would generate an added $750 over 25 years. The impact of shaded parking includes a reduced temperature in the car and less loss of value when selling or trading in the vehicle. If there were an added charge of $250 each year for the convenience and opportunity to use a charge station and purchase electricity from the local utility company at 7 cents per kWh, this would generate $6250 of income in 25 years. One can also consider the view of a BEV owner who commutes 40 or 60 miles to work at the university, which would consume 13.3 kWh and 20 kWh of energy, respectively. When they arrive on campus, they can use the SPCS to charge their vehicle while they work. If they paid for the $250 permit to use the station and assuming they work about 250 days each year, they would pay about $1 per day to use the station. Furthermore, the cost to charge at the university would be 7 cents per kWh and 12.12 cents per kWh at home, making the total cost to charge about $2.55 and $3.82 each day for a 40- and 60-mile commute, respectively. This is much cheaper than the fuel cost of driving the average vehicle in 2012, which would amount to $13.85 and $20.77 each day for a 40- and 60-mile commute, respectively.

If the current pricing system for charging using this university's existing Level 2 charger located in the parking garage ($3 for the first hour and $1 for each additional hour) was used for the SPCS, rather than paying for a permit and the price of electricity, the costs would increase. The charge station delivers about 7.2 kW of power, so two hours of charging would be needed to recharge the battery after a 40-mile commute, and three hours of charging after a 60-mile commute. The cost would add on $4 for the 40-mile commute each day and $5 to the 60-mile commute, bringing the total prices to $6.65 each day for a 40-mile commute and $7.42 each day for a 60-mile commute, assuming the price to charge from home is still $0.1212 per kWh.

The amount of energy that could potentially be produced by installing SPCSs nationwide is also something to consider when looking at the economics of the stations. To calculate this, we assume an installation of 100 million, 150 million, and 200 million stations across the United States, with installation costs of $10,000, $15,000, and $20,000 per station and an average energy price of 10.08 cents per kilowatt hour. These are fairly realistic scenarios in terms of logistics, as there are an estimated 750 million parking spots in the United States where these stations could be installed (Chester et al., 2010). Tables 9.6 and 9.7 show the results of these calculations. The cost to install these stations would be very great, from $1 trillion to $4 trillion, depending on the number of stations installed and their installation costs. They would also produce a large amount of energy, from about 580–1160 TWh per year, which is about 14%–29% of the amount of energy generated in 2013. However, as stated earlier, the real measure of economic viability is if the stations can pay for themselves before needing replacement. In this case, an installation cost of $15,000 will be paid back in just over 25 years, so any installation costs higher than this would not be viable. This would, however,

TABLE 9.6

Percentage of Total Electrical Energy that Could Be Produced Through Solar Powered Charge Stations Assuming Each Charging Station Produces 16 kWh of Energy Each Day for 365.25 Days per Year Based on 4058 TWh of Electrical Energy Produced in 2013

Number of Solar Powered Charge Stations Installed (Millions)	Energy Produced from Stations Each Year (TWh)	Percentage of Energy Produced by SPCSs in Relation with Energy Produced in 2013
100	584.4	14%
150	876.6	22%
200	1169	29%

Source: U.S. Energy Information Administration. 2014e. Monthly energy review. Retrieved May 28, 2014 from http://www.eia.gov/totalenergy/data/monthly/.

TABLE 9.7

The Effect of SPCSs on Energy Generation and Greenhouse Gas Emissions Based on 2775 TWh of Electricity Produced from Fossil Fuels in 2013

Number of Solar Powered Charge Stations Installed (Millions)	Solar Energy Produced Each Year (TWh)	Equivalent Fraction of Energy Produced by Burning Fossil Fuels in 2012	Tg of CO_2e Emissions Eliminated Annually from Reduced Fossil Fuel Energy Generation Mix
100	584.4	21.06%	426.0
150	876.6	31.59%	639.0
200	1169	42.12%	851.9

Source: U.S. Energy Information Administration. 2014e. Monthly energy review. Retrieved May 28, 2014 from http://www.eia.gov/totalenergy/data/monthly/; United States Environmental Protection Agency. 2014. Inventory of U.S. greenhouse gas emissions and sinks: 1990–2012. (No. EPA 430-R-14-003).

create many jobs for American citizens to ship, manufacture, install, and maintain all of the supplies necessary to install the stations. Additionally, this would have a huge impact on the annual amount of greenhouse gas emissions released into the atmosphere. Table 9.7 outlines the impact of the SPCSs on the amount of greenhouse gases released into the atmosphere from the electricity-generation sector based on 2012 data, where 2775 TWh of electricity was produced from all fossil fuel sources in the United States. If 200 million charge stations were installed in the United States, greenhouse gas emissions produced from the national mix of fossil-fuel energy generation could be reduced by as much as 42%. The values in Tables 9.6 and 9.7 may be compared to 286 TWh per year of electrical energy that would be used by a fleet of 40 million BEVs and 60 million PHEVs, each driving 11,300 miles per year with an efficiency of 3 miles per kWh, and with 60% of the PHEV miles being powered by electricity.

Increased electricity generation, greater infrastructure for recharging electric vehicles, and providing shade for parked cars will help increase the adoption of electric vehicles while decreasing emissions of greenhouse gases and other toxins. Although SPCSs are too expensive to install in some sectors, the prices of solar panels themselves are expected to continue to fall as the technology becomes more developed. Improvements in the efficiency of the solar panels are also expected. Furthermore, it will eventually become less expensive than fossil fuels, since those prices are expected to continue to increase (Chiras, 2012). In addition to these direct benefits, there are some indirect benefits to installing an SPCS as well. For example, the installation of these stations would make a statement about the home or business installing them. Just as many companies already advertise their products as being "green" to make a statement about the product's relationship to the environment, companies who install SPCSs will be making a statement like this as well. This has been observed in communities like California, where using "green" products is a sort of status symbol and actually drives their adoption—dubbed the "Prius Effect" (UC Berkeley and UCLA release "the value of green labels in the California housing market" study findings, 2012). This would attract more consumers to a business with SPCSs installed in the parking lots, providing an even greater economic boost from the stations. Another way SPCSs could be used to attract customers is to use them as an advertising space for companies. The company Volta has done this with EV charging stations in places like shopping malls and grocery stores. The stations are placed close to store entrances to give EV drivers preferential parking. Furthermore, Volta sells advertising space on the stations themselves to cover the cost of installation and maintenance, as well as to offer the charging service for free. According to Volta's website, retailers have seen that customers enjoy the charging stations and choose their stores over others for the opportunity to charge for free (Volta, 2015). This presents a unique opportunity for business owners when considering powering the charging stations through solar power. Not only is the charging station paid for through advertisements, but the solar panels can provide electricity to the business when the charging station is not in use. The solar panels could also potentially provide more advertising space to be sold.

9.6 Financing and Policy

There are many options available to finance the installation and maintenance of SPCSs. A survey conducted by Robinson et al. (2014) showed that several different financing options such as federal grants and partnerships with electric charging station companies would be well received by the public (Robinson et al., 2014).

Federal and state grants have been traditionally used to finance the installation of SPCSs (Goldin et al., 2014). There are many state and federal policies that apply to EVSEs, mostly involving rebates for their installation costs. In 2008, Washington and Oregon made plans to increase the number of charging stations along Interstate 5 to facilitate electric vehicle travel. At the end of 2014, there were charging stations located every 25–50 miles along Interstate 5, which accounted for about 9% of all public charging stations in the United States at the time (McFarland and Chase, 2014). This can be partially attributed to incentives provided to businesses for installing such charge stations. As discussed previously, charging stations at retail stores and other public areas can attract customers. Furthermore, Oregon offers a tax credit for the installation of the charge stations of 35% of the installation cost for businesses (McFarland and Chase, 2014).

Power purchase agreements (PPAs) have also been increasingly utilized to finance SPCS installation. A PPA allows a customer to purchase the electricity produced by a solar power system at a reduced cost without purchasing the system itself—similar to a lease on a car (U.S. Department of Energy, 2014b). After the agreement ends, the buyer may either renew the agreement or purchase the system (U.S. Department of Energy, 2014b). This makes it easier for businesses to purchase and provide SPCSs for their customers, since they only have to pay for the electricity produced. However, many states have disallowed power purchase agreements or are unclear on their regulation (U.S. Department of Energy, 2014b).

Income from customers using the charging station is also a viable option to offset the cost of SPCS installation. In 2012, the University of California, San Diego installed charging stations on their campus (Uda, 2012). In order to use the stations, users must have purchased a parking permit with the university or paid for visitor parking (Uda, 2012). Users also must pay for the electricity they consume from the station, amounting to about $3 for an 80% charge (Uda, 2012). With the shade SPCSs provide, users could also be charged for the privilege to park underneath the panels, providing another source of income. One popular example of charging station financing is the Tesla Supercharger network, in which the charging stations are financed by the purchase of Tesla vehicles and used solely by those vehicle owners (Robinson et al., 2014).

Time-of-use (TOU) pricing can also be a factor for financing of SPCSs. Availability for TOU pricing is becoming more common as smart grid programs develop. For instance, Westar Energy, an electric utility company in Kansas, is seeking customers to volunteer for their TOU rate pilot project (Westar Energy, 2014). Other electric utility companies, like Portland General Electric (PGE) in Oregon, have already developed and instituted TOU pricing options for their customers (Portland General Electric, 2015). PGE provides electricity at different rates for on-peak, mid-peak, and off-peak hours, and clearly defines what times fall into which categories (Portland General Electric, 2015). However, both Westar and PGE stress that while TOU pricing is meant to offer an opportunity for customers to reduce their electric bills, there is a possibility that if too much electricity is consumed during on-peak hours, the customer may actually see their electric bill increase (Portland General Electric, 2015; Westar Energy, 2014). This is generally good news for EV owners, as they are able to use their vehicles during the day (on-peak hours) and charge them at night (off-peak hours) when electricity is cheap. However, sometimes charging during on-peak hours can be unavoidable. This is where an SPCS would be especially useful. Since the electricity from the SPCS would not be coming directly from the grid during on-peak hours, the customer can still avoid charging their vehicle with the higher on-peak rates. TOU pricing increases the value of the electricity from the solar panels because prices are higher at mid-day, when solar panels are receiving the most solar radiation.

Recently, Ye et al. (2015) have completed a feasibility study of SPCSs with results that illustrate the importance of TOU prices on the economics of an integrated system of EVs, SPCSs, and the electrical grid. They report the cost of energy of the combined system is $0.098/kWh.

9.7 Relation to Sustainable Development

The adoption of electric vehicles and of SPCSs is good practice in sustainable development. In 2012, the Rio +20 conference hosted by the United Nations created an outcome document entitled "The Future We Want" (The United Nations, 2012). The document outlines several goals for sustainable development, many of which can be aided by the adoption of electric vehicles and SPCSs. In terms of energy and transportation, the document declares:

> Improving energy efficiency, increasing the share of renewable energy, cleaner and energy-efficient technologies are important for sustainable development, including in addressing climate change. We also recognize the need for energy efficiency measures in urban planning, buildings, and transportation [...] Sustainable transportation can enhance economic growth as well as improving accessibility. Sustainable transport achieves better integration of the economy while respecting the environment. (The United Nations, 2012)

Solar powered charge stations and electric vehicles satisfy both of these claims to increase sustainable development. The SPCSs produce clean, renewable energy that the electric vehicles then consume, producing zero emissions. This results in reduced consumption of fossil fuels for transportation and energy generation, creating a more sustainable infrastructure. The U.N. also published another document in 1992 entitled Agenda 21, which outlines specific goals for governments of all nations to encourage sustainable development. The document states that "Governments should explore, in cooperation with business and industry, as appropriate, how effective use can be made of economic instruments and market mechanisms in [...] issues related to energy, transportation" (The United Nations, 1992). This is very important in relation to the adoption of EVs and SPCSs. Currently, both have high initial costs but have long-term economic benefits. Government subsidies and other benefits are vital to encourage the adoption of these technologies by a vast majority of consumers in order to make the technologies effective and further the practice of sustainable development. Although emphasis in this chapter is on applications for the United States, these ideas have value in many parts of the world.

9.8 Conclusions

Electric vehicles and SPCSs, while expensive in the current market, are extremely cost-effective forms of energy production and transportation. The benefits each of these technologies has on human health, climate change, the environment, and foreign affairs are great. Not only do they offer small long-term operational costs, they also produce no emissions by themselves and require less maintenance than most other competing technologies. Moreover, they also help meet the sustainable development goals that have been set forth by the United Nations. Overall, as the technology develops, we can expect EVs and SPCSs to make a significant impact around the world. The proposed advances can be accomplished with our current battery, EV, and solar power technologies, but expected innovations in technology efficiency will expedite the transition.

References

Benbrahim-Tallaa, L., Baan, R. A., Grosse, Y., Lauby-Secretan, B., El Ghissassi, F., Bouvard, V., and Straif, K. 2012. Carcinogenicity of diesel-engine and gasoline-engine exhausts and some nitroarenes. *Lancet Oncology*, 13(7): 663. doi:10.1016/S1470-2045(12)70280-2.

Bullis, K. Jan/Feb 2013. How improved batteries will make electric vehicles competitive. *Technology Review*, 116: 19–20.

Chester, M. , Horvath, A., and Madanat, S. 2010. Parking infrastructure: Energy, emissions, and automobile life-cycle environmental accounting. *Environmental Research Letters*, 5(3): 034001. doi:10.1088/1748-9326/5/3/034001.

Chiras, D. Apr/May 2012. Cost of solar energy plummets. *Mother Earth News*, 14.

Davis, S. C., Diegel, S. W., and Boundy, R. G. 2013. *Transportation Energy Data Book*, 32nd ed. Oak Ridge, TN: Oak Ridge National Laboratory.

Davis, T. 2014. Retrieved July 2014 from http://www.k-state.edu/media/newsreleases/jan13/station13013.html.

Electric Drive Transportation Association. 2014. Electric drive sales. Retrieved June 2014 from http://electricdrive.org/index.php?ht=d/sp/i/20952/pid/20952.

Erickson, L. E., Boguski, T., Babcock, M. W., Leven, B. A., Pahwa, A., Brase, G. L., and Miller, R. D. 2012. Community infrastructure for the electrification of transportation. Retrieved from https://www.engg.ksu.edu/chsr/.

ExxonMobil. 2014. The outlook for energy: A view to 2040. ExxonMobil. Retrieved from http://cdn.exxonmobil.com/~/media/Reports/Outlook%20For%20Energy/2014/2014-Outlook-for-Energy-print-resolution.pdf.

Goldin, E., Erickson, L., Natarajan, B., Brase, G., and Pahwa, A. 2014. Solar powered charge stations for electric vehicles. *Environmental Progress & Sustainable Energy*, 33: 1248–1308. doi:10.1002/ep.11898.

Goli, P. and Shireen, W. 2014. PV powered smart charging station for PHEVs. *Renewable Energy*, 66: 280–287. doi: http://dx.doi.org/10.1016/j.renene.2013.11.066.

GTM. 2015. U.S. Solar Market Insight Report, Q3, Executive Summary, GTM Research, 2015; http://www.greentechmedia.com/.

Hargreaves, S. 2012. Gas prices hit highest average ever in 2012. CNN Money, December 31, 2012.

Horn, M. and Docksai, R. 2010. Roadmap to the electric car economy. *The Futurist*, 44(2): 40–45.

Hunter, H. 2012. Cost of owning and operating vehicle in U.S. increased 1.9 percent according to AAA's 2012 "your driving costs" study. Retrieved from http://newsroom.aaa.com/2012/04/cost-of-owning-and-operating-vehicle-in-u-s-increased-1-9-percent-according-to-aaa%E2%80%99s-2012-%E2%80%98your-driving-costs%E2%80%99-study/.

In 2012, the market for REDD carbon credits shrank and the price fell. Retrieved June 2014 from http://www.redd-monitor.org/2013/07/17/in-2012-the-market-for-redd-carbon-credits-shrank-and-the-price-fell/.

ISO New England Inc. 2003. Hourly historical data post-market 2002. Retrieved from http://www.iso-ne.com/markets/hstdata/hourly/his_data_post/index.html.

Jamil, M., Kirmani, S., and Rizwan, M. 2012. Techno-economic feasibility analysis of solar photovoltaic power generation: A review. *Smart Grid and Renewable Energy*, 3(4): 266–274.

Markandya, A., and Wilkinson, P. 2007. Energy and health 2: Electricity generation and health. *The Lancet*, 370(9591): 979–990.

McFarland, A. and Chase, N. 2014. Several states are adding or increasing incentives for electric vehicle charging stations. Retrieved from http://www.eia.gov/todayinenergy/detail.cfm?id=19151.

Nykvist, B. and Nilsson, M. 2015. Rapidly falling costs of battery packs for electric vehicles. *Nature Climate Change*, 5(4): 329–332. doi:10.1038/nclimate2564.

Plumer, B. 2014. A guide to Obama's new rules to cut carbon emissions from power plants. *Vox*. Retrieved from http://www.vox.com/2014/6/1/5770556/EPA-power-plant-rules-explainer.

Portland General Electric. 2015. Time of use: Pricing | PGE. Retrieved from https://www.portlandgeneral.com/residential/your_account/billing_payment/time_of_use/pricing.aspx.
Ramsey, M. 2010. Plugging in: As electric vehicles arrive, firms see payback in trucks. *Wall Street Journal*, Dec. 8, 2010.
Ramsey, M. 2014. Why electric cars click for Atlanta. *The Wall Street Journal*, June 5, 2014, p. B8.
Robinson, J., Brase, G., Griswold, W., Jackson, C., and Erickson, L. 2014. Business models for solar powered charging stations to develop infrastructure for electric vehicles. *Sustainability*, 6(10): 7358–7387.
Rosen, L. 2014. New aluminum-air battery powered car travels 1,800 miles without a recharge. Retrieved from http://www.wfs.org/blogs/len-rosen/new-aluminum-air-battery-powered-car-travels-1800-kilometers-without-recharge.
Small, K. A. and Kazimi, C. 1995. On the costs of air pollution from motor vehicles. *Journal of Transport Economics and Policy*, 29(1): 7.
Smith, D. 2014. Elon Musk may give away its Tesla Supercharger patents to spur electric car development. *Business Insider*, June 9, 2014.
Smith, K. R., Jerrett, M., Anderson, H. R., Burnett, R. T., Stone, V., Derwent, R., and Thurston, G. 2009. Health and climate change 5: Public health benefits of strategies to reduce greenhouse-gas emissions: Health implications of short-lived greenhouse pollutants. *The Lancet*, 374(9707): 2091–2103.
The United Nations. 1992. Agenda 21. United Nations Conference on Environment and Development, Rio de Janero, Brazil.
The United Nations. 2012. The future we want. (No. A/RES/66/288). The United Nations.
Tulpule, P., Marano, V., Yurkovich, S., and Rizzoni, G. 2013. Economic and environmental impacts of a PV powered workplace parking garage charging station. *Applied Energy*, 108, 323–332. doi: http://dx.doi.org/10.1016/j.apenergy.2013.02.068.
U.S. Department of Energy. 2014a. Retrieved from http://www.fueleconomy.org.
U.S. Department of Energy. 2014b. Third party solar financing. Retrieved from http://apps3.eere.energy.gov/greenpower/onsite/solar_financing.shtml.
U.S. Energy Information Administration. 2011. Direct federal financial interventions and subsidies in energy in fiscal year 2010. Retrieved June 2014 from http://www.eia.gov/analysis/requests/subsidy/.
U.S. Energy Information Administration. 2013. How dependent are we on foreign oil? Retrieved from http://www.eia.gov/energy_in_brief/article/foreign_oil_dependence.cfm.
U.S. Energy Information Administration. 2014a. Electric power monthly, Table 1.1. Net Generation by Energy Source: Total (all sectors), 2004–March 2014. Retrieved May 2014 from http://www.eia.gov/electricity/monthly/index.cfm.
U.S. Energy Information Administration. 2014b. U.S. refinery yield. Retrieved from http://www.eia.gov/dnav/pet/pet_pnp_pct_dc_nus_pct_a.htm.
U.S. Energy Information Administration. 2014c. Annual energy outlook 2014, table 58. (No. DOE/EIA-0383(2014)).
U.S. Energy Information Administration. 2014d. Average retail price of electricity to ultimate customers. Retrieved from https://www.eia.gov/electricity/monthly/epm_table_grapher.cfm?t=epmt_5_3.
U.S. Energy Information Administration. 2014e. Monthly energy review. Retrieved May 28, 2014 from http://www.eia.gov/totalenergy/data/monthly/.

U.S. Energy Information Administration. 2014f. Product supplied for finished motor gasoline. Retrieved May 29, 2014 from http://www.eia.gov/dnav/pet/hist/LeafHandler.ashx?n=PET&s=MGFUPUS1&f=M.

UC Berkeley and UCLA release "the value of green labels in the California housing market" study findings. 2012. *Health & Beauty Close-Up*, n/a. Retrieved from http://search.proquest.com/docview/1027615531?accountid=11789.

Uda, R. 2012. UCSD installs first electric vehicle charging station. *The Guardian*. Retrieved from http://ucsdguardian.org/2012/10/8/ucsd-installs-first-electric-vehicle-charging-stations/.

United States Environmental Protection Agency. 2014. Inventory of U.S. greenhouse gas emissions and sinks: 1990–2012. (No. EPA 430-R-14-003).

Up to 90% of US cars could be "green" vehicles by 2030. 2011. *Business and the Environment*, 22(10): 8.

Victoria Transport Policy Institute. 2009. Transportation cost and benefit analysis. Retrieved from http://www.vtpi.org/tca/.

Volta. 2015. Volta charging. Retrieved from http://www.voltacharging.com/home.

Waco, D. 2011. How long do solar panels last? Retrieved from http://www.civicsolar.com/resource/how-long-do-solar-panels-last.

Wadud, Z. 2011. Comparison of air quality-related mortality impacts of different transportation modes in the United States. *Transportation Research Record*, 2233: 99–109. doi:10.3141/2233-12.

Westar Energy. 2014. Time of use rate—Voluntary pilot program. Retrieved from https://www.westarenergy.com/time-of-use-rate.

Ye, B., Jiang, J., Miao, L., Yang, P., Li, J., and Shen, B. 2015. Feasibility study of a solar-powered electric vehicle charging station model, *Energies*, 8: 13265–13283.

Zhang, L., Brown, T., and Samuelsen, S. 2013. Evaluation of charging infrastructure requirements and operating costs for plug-in electric vehicles. *Journal of Power Sources*, 240: 515–524.

10

Sustainable Development

Larry E. Erickson, Jessica Robinson, Jackson Cutsor, and Gary Brase

CONTENTS

One planet, one experiment.

Edward O. Wilson

10.1 Introduction

Sustainable development is an important topic for this book. The infrastructure of solar powered charging stations (SPCSs) is one small part of sustainable development. In this chapter, the topics of sustainable energy, transportation, smart grid, and SPCSs are considered with an emphasis on sustainable development.

10.2 Paris Agreement on Climate Change

On December 12, 2015, the Paris Agreement on Climate Change was adopted by the Parties to the United Nations Framework Convention on Climate Change. The agreement aims to strengthen the global response to the threat of climate change, in the context of sustainable development and efforts to eradicate poverty. The goal is to hold the increase in global average temperature to less than 2°C above preindustrial levels and to pursue efforts to limit the increase to 1.5°C. The parties aim to reach global peaking of greenhouse gas emissions as soon as possible and a balance between anthropogenic emissions and removals by sinks of greenhouse gases in the second half of this century (UNFCCC, 2015).

The Paris Agreement on Climate Change is important because the goals related to the increase in temperature associated with greenhouse gas emissions are stated clearly, and there is strong support from many countries to accomplish the goals of the agreement. The social mobilization that is important to accomplish the goals in each country can move forward with greater participation because there has been agreement on the global climate change goals. Many citizens will draw hope from the agreement and take action to help reduce greenhouse gas emissions.

The goal to reach a point where concentrations of greenhouse gases in the atmosphere stop increasing in this century is a recognition of the importance of climate change and the urgency of taking action. Transforming our societies and daily activities to achieve a balance between emissions and removals is the greatest challenge that the United Nations and the citizens of the world have faced.

The electrification of transportation and the generation of electricity with solar and wind energy are examples of what we need to do to achieve the goals of the Paris Agreement. The installation of SPCSs in many parking lots in all parts of the world can help many countries make progress toward their goals for the Paris Agreement.

10.3 United Nations Sustainable Development Goals

Earlier in 2015, the United Nations Sustainable Development Goals were adopted (United Nations, 2015). The 17 sustainable development goals are as follows (United Nations, 2015):

1. End poverty in all its forms everywhere;
2. End hunger, achieve food security and improved nutrition and promote sustainable agriculture;
3. Ensure healthy lives and promote well-being for all at all ages;
4. Ensure inclusive and equitable quality education and promote lifelong learning opportunities for all;
5. Achieve gender equality and empower all women and girls;
6. Ensure availability and sustainable management of water and sanitation for all;
7. Ensure access to affordable, reliable, sustainable and modern energy for all;

8. Promote sustained, inclusive, and sustainable economic growth, full and productive employment and decent work for all;
9. Build resilient infrastructure, promote inclusive and sustainable industrialization and foster innovation;
10. Reduce inequality within and among countries;
11. Make cities and human settlements inclusive, safe, resilient and sustainable;
12. Ensure sustainable consumption and production patterns;
13. Take urgent action to combat climate change and its impacts;
14. Conserve and sustainably use the oceans, seas, and marine resources for sustainable development;
15. Protect, restore, and promote sustainable use of terrestrial ecosystems, sustainably manage forests, combat desertification, and halt and reverse land degradation and halt biodiversity loss;
16. Promote peaceful and inclusive societies for sustainable development, provide access to justice for all and build effective, accountable and inclusive institutions at all levels;
17. Strengthen the means of implementation and revitalize the global partnership for sustainable development.

We can relate some of these goals to topics in this book. Goal 2 includes sustainable agriculture, which must include a transition to sustainable energy. The electrification of farming with sustainable energy can include SPCSs for use with farm tractors and other equipment that is powered by electricity.

Goal 3 includes good health, which connects with Chapter 8 and the air quality benefits from electrification of transportation in urban areas. The well-being of many people can be enhanced by providing shaded parking under solar panels and better air quality in cities.

The challenge of having sufficient sustainable energy for all (Goal 7) can be met by using SPCSs in remote areas as a source of electricity for off-grid home needs as well as for transportation. Solar powered charge stations and battery storage can provide evening lighting and cell phone charging in off-grid locations. To support the Paris Agreement, SPCSs are needed to provide electricity to the grid and to EV batteries.

If we move forward with the transition to wind and solar energy for electric power and SPCSs for EVs, many people will be employed in the process of accomplishing the sustainable economic growth goals described in Goal 8.

Goal 9 includes building resilient infrastructure and sustainable industrialization. The SPCSs, EVs, and wind and solar energy fit here, also. We can also expect that the transition to these technologies will foster further innovation.

Goal 11 is to make cities and human settlements sustainable. Clean air, EVs, SPCSs, sustainable energy, good walking conditions, bike paths, and green spaces can contribute to this goal. Solar panels to generate electricity on buildings and in parking lots capture solar energy and produce electricity. This helps to cool the urban environment on hot summer days.

Goal 12 is to have sustainable production and consumption, which means that recycling must be included in the transition to EVs and SPCSs. Solar panel recycling, battery recycling, and EV recycling must become standard practices. Solar panels, EVs, and batteries must be made out of raw materials that will be available to future generations, also.

Goal 13 relates to reducing greenhouse gas emissions, which is advanced by SPCSs and EVs.

10.4 Complexity of Sustainable Development

The challenge of following a path that leads to sustainable development has many dimensions to it, and there are many complex systems that are difficult to optimize. Sachs (2015) addresses many of these important complex topics and issues. His book includes significant information on many aspects of sustainable development, which we do not include here, and we encourage those interested in sustainable development to read this excellent book. The Paris Agreement provides some specifics and details on how countries plan to work together to address sustainable development of Goal 13 to combat climate change and its impacts (UNFCCC, 2015).

One of the greatest challenges related to reducing greenhouse gas emissions is to develop science and technology that makes SPCSs, EVs, and battery storage so inexpensive that companies and consumers find them the best choice when they make decisions. New developments in solar panels, batteries, and EVs have already changed the world, and there is much ongoing research. Solar energy is already a great success in Germany, and from January to September 2015, 30% of all new electric generating capacity brought online in the United States was solar (GTM, 2015). Seba (2010, 2014) expects solar energy to become a trillion dollar industry, and by 2030 Seba expects EVs and renewable energy to take over and be the most attractive options. The costs for solar panels and batteries are expected to decrease to the point where EVs will be much less expensive compared to other vehicles, and electricity from solar will be very low cost and efficient. Developments in smart grid systems technology will allow EVs and SPCSs to be an integral part of the grid. Because of the trillions of dollars associated with the manufacture and installation of solar panels and batteries, there are opportunities in this area for both academic research and industrial research and development in many countries of the world.

In short, there are great opportunities within the upcoming transitions in renewable electric power with smart grids, SPCSs, energy storage, and the electrification of transportation with EVs and SPCSs. There will also be many investment and employment opportunities. This will change life for all people everywhere.

The roles of government and policy in sustainable development are very important. The Paris Agreement is significant in that it demonstrates leaders' commitments to acting on their discussions. It is necessary to take additional steps to accomplish the goals of the agreement to keep temperature increases well below 2°C. Government support for research on solar panels and batteries has been an important part of the progress to reduce the cost and improve the quality of these products. Government support for research on renewable energy, smart grid, batteries, EVs, and SPCSs should continue across the countries of the world.

Government incentives to encourage progress in renewable energy and electrification of transportation have been helpful. With the transition to renewable solar and wind power generation, policies are needed that continue to encourage progress. The financial incentives that have been available for the installation of wind and solar power generation systems have been crucial for supporting this budding technology. Because of the importance of inexpensive energy to the export of products and international trade, financial support to encourage wind and solar production of electricity has been easier to accomplish than imposing a carbon tax on greenhouse gas emissions.

There is also the need to address current and potential barriers to progress. For example, the integration of solar panels on homes and businesses into the grid at a fair price for power is important. There are both opportunities and challenges for electric utility companies in this great transition to renewable power because of the importance of the smart grid, SPCSs, and reliability. With the electric power smart grid infrastructure including renewable generation, SPCSs, energy storage, EV charging, and more active participation of consumers with time of use (TOU) prices, there will be important policy issues that will need to be negotiated and approved. The organizations that regulate utilities have new challenges and responsibilities because of distributed renewable generation, utilities selling power to charge EVs in parking lots with SPCSs, utilities selling advertising at SPCSs, and the agreements between utilities and owners of properties with solar panels that generate power that flows into the grid.

It is important to have the participation of governments, industry, and other organizations in the processes of electrifying transportation moving forward with the smart grid and generating electricity without greenhouse gas emissions. Financial support from foundations, the World Bank, and charitable organizations has been valuable in past sustainable development projects, and financial resources will be needed to make the transition to SPCSs, EVs, and a smart grid with renewable energy and energy storage.

Sachs (2015) addresses the importance of good governance in accomplishing sustainable development goals. The rules of behavior in corporations, foundations, units of government, NGOs, and other organizations should reflect good governance. Large multinational corporations have significant power and their behavior and actions are important. Specifically, accountability is one of the principles of good governance. If all organizations have goals that are supportive of the sustainable development goals, take actions to accomplish their goals, report on their progress and provide transparency, this is an example of being accountable. There should be active participation by many different stakeholders such that good ideas are encouraged and implemented in these organizations. In our modern world, multidisciplinary teams are often able to accomplish tasks effectively. Diversity that is effectively managed often leads to significant advances that have value for society.

The Paris Agreement addresses one of the important economic externalities—greenhouse gas emissions. One of the principles of good governance is to address environmental impacts so that organizations do no harm; that is, they do not have significant environmental degradation associated with their activities and operations. The ethical responsibility to do no harm is applicable to both individuals and organizations.

Many organizations have made a commitment to sustainable development and produce reports on progress to address efforts to follow a pathway toward sustainability. The commitment by organizations to follow the principles of good governance and support efforts to accomplish the goals of the Paris Agreement will be very beneficial.

Urban environments in medium and large cities have many challenges to transition from their present states to being carbon neutral such that there are no net greenhouse gas emissions. Parking lots with SPCSs, solar panels on buildings, public electric transportation, electric heating and air conditioning, EVs, and electric bicycles will be features of these modern sustainable urban environments. The quality of life in cities can be improved through urban planning that brings the urban environment closer to the pathway to a sustainable city.

The U.N. sustainable development Goal 4 is to have quality education for everyone. Education is one of the most significant aspects of sustainable development because it impacts progress toward many of the other sustainable development goals, including the Paris Agreement. Education is needed for the smart grid to function effectively with customers responding appropriately to TOU prices for electricity. Education will be needed in those parts of the world where solar panels with battery storage can bring inexpensive power to those who have not had power because there is no grid nearby. With inexpensive solar power and inexpensive battery storage, sustainable development Goal 7 can be achieved (access to affordable, reliable, sustainable, and modern energy for all). However, those who have their own power system will need to understand its design and operation so they can take care of it and maintain it.

One of the reasons to have an extensive educational system is to educate as many people as possible with the training that they need to be active participants in the process to accomplish tasks in support of the Paris agreement and sustainable development goals. If electric power can be provided to all people everywhere, one can provide education over the Internet and other electronic networks to everyone. With education for everyone, people can move toward making more informed, responsible, and sustainable decisions in their lives. A virtuous cycle is created.

There is a need to have solid support systems when decisions related to sustainable development need to be made. Social value, environmental impacts, and economics should be included in the development of decision support systems. Goldin et al. (2014) illustrate how to consider these triple bottom line values in developing SPCSs for EVs. Decision support systems can be integrated into the smart grid such that some decisions are automated.

10.5 Conclusions

The Paris Agreement on Climate Change is a giant step forward. The agreement is beneficial to the mobilization of individuals, industry, governments, and other organizations to work cooperatively to reduce greenhouse gas emissions while working to accomplish the sustainable development goals. Progress in sustainable development will happen in this century to make SPCSs and EVs more attractive and competitive so that hundreds of millions of them are in daily use. The transition to EVs being the vehicle drivetrain of choice will change the world significantly. It is part of the transition to electrify transportation and generate electricity without greenhouse gas emissions. We have already started on this transition. With advances in science, technology, prices, new product availability, and consumer behavior, the rate of this transition will accelerate. Education will allow more people to be part of this great transition.

References

Goldin, E., L. Erickson, B. Natarajan, G. Brase, and A. Pahwa. 2014. Solar powered charge stations for electric vehicles, *Environmental Progress and Sustainable Energy*, 33: 1298–1308.

GTM. 2015. U.S. Solar Market Insight, GTM Research, SEIA; http://www.gtmresearch.com.

Sachs, J.D. 2015. *The Age of Sustainable Development*, Columbia University Press, New York.
Seba, T. 2010. *Solar Trillions*, Seba Group, San Francisco, CA.
Seba, T. 2014. *Clean Disruption of Energy and Transportation*, Clean Planet Ventures, Silicon Valley, CA.
United Nations. 2015. Transforming our world: The 2030 agenda for sustainable development, UN Division for Sustainable Development; https://sustainabledevelopment.un.org/.
UNFCCC. 2015. Paris Agreement, United Nations Framework Convention on Climate Change, FCCC/CP/2015/L.9, December 12, 2015; http://unfccc.int/.

11

International Opportunities

Jessica Robinson, Larry E. Erickson, and Jackson Cutsor

CONTENTS

The moral imperative to make big changes is inescapable...That what we take for granted may not be here for our children.

Al Gore

By 2050 there are expected to be 9 to 10 billion people on Earth. That is around 2 billion more than in 2015. This drastic increase in human population has many implications associated with it, but increased carbon emissions are one important facet. With the population of developing countries growing exponentially, there will be greater stress on all resources, specifically energy. There are also projections of the last massive human migration in history. This migration is from rural settlements to urban environments throughout the world. With a larger percentage of the population being in urban environments, there will no doubt be greater demands for electricity and cars. This increased cultural carrying capacity brings new obstacles along with many new opportunities to solve them. Many European countries are leading the way with green technology and power generation. Increasing EV adoption will help meet the December 2015 Paris Agreement, which over 190 countries committed to in order to combat climate change.

The electric vehicle industry is constantly progressing. It is important to understand the current development of EVs and charge station infrastructure worldwide. This chapter considers the EV situation in leading countries in Europe, Asia, and Oceania. More specifically, this chapter considers the current status of EVs, policies, incentives, charge station infrastructure, and issues and potential improvements for Norway, the Netherlands, the United Kingdom, France, Germany, Denmark, China, Japan, and Australia.

By the end of 2014, less than 1% of all passenger cars sold worldwide were EVs. The majority of money spent on EVs has been in the areas of research, development, and demonstration, followed by fiscal incentives for EV purchases, with less money invested in charge station infrastructure. There are over 15,000 fast charging points installed and over 94,000 slow charging points worldwide. The United States had the greatest share of the global EV stock in 2014 with 39% followed by Japan with 16% and China with 12% (Global EV Outlook 2015, 2014).

The Electric Vehicles Initiative (EVI) is an international policy forum devoted to increasing worldwide adoption of EVs. EVI members include 16 governments from countries in Africa, Asia, Europe, and North America, and the International Energy Agency. Over 95% of the global EV stock is in EVI countries (Global EV Outlook 2015, 2014). The data values for countries' EV and EVSE stocks and policies and incentives listed are the most recent information available as of June 2015.

11.1 EV Sales Worldwide

How do electric vehicle sales compare across Europe, North America, and China? In the first half of 2015, Europe led in EV sales with over 78,000 sales; 34,000 more than China and 25,000 more than North America. Just in the month of June 2015, Europe sold about 16,000 EVs while China sold 10,000 and North America sold 9750 EVs (Zach, 2015). Europe is currently the world leader for EV sales. When comparing countries' total EV stock, though, the United States is far in the lead with 275,000 EVs, followed by Japan with 108,250 EVs, and China with 83,200 EVs. When comparing total public charge station stock using 2014 data, China is leading with 30,000 charge stations, followed by the United States with 21,800 charge stations, and the

Netherlands with 12,114 charge stations (Global EV Outlook 2015, 2014). Of course, it is difficult to simply compare EV and EVSE stock values to determine the leading countries for electric vehicles since countries have varying population sizes, landmasses, and geographies. It does appear, however, that every global region has strengths and challenges to work on.

11.2 Europe

11.2.1 Top EV Car Sales

The top three vehicles sold in Europe, using data for June 2015, are the Mitsubishi Outlander PHEV, the Nissan Leaf, and Renault Zoe (EV Sale, 2015). The Mitsubishi Outlander PHEV is larger than the average EV. It has 5 seats and it is 15 feet by 5 feet. Also, the Renault Zoe is a relatively inexpensive EV starting around $20,480 with a spacious cabin. The top European leaders in EV sales are Norway and the Netherlands followed by the United Kingdom, France, and Germany (EVolution, 2014).

11.2.2 Policy

The European Union has short-term and long-term carbon emission targets for light vehicles. As of 2012, the EU announced a target of 130 g CO_2/km, for 2020 a target of 95 g CO_2/km, and for 2025 a potential target of 68–78 g CO_2/km, which will be confirmed in 2016. In order to meet the 2020 target, OEMs (original equipment manufacturers) will need to reduce their vehicle carbon emissions by about 30%, likely leading to developments in electrified transportation (EVolution, 2014). In order to encourage EV adoption rates, EU governments are investing in EV infrastructure and providing incentives for drivers such as subsidies, tax breaks, and special driving privileges.

Electric vehicle car sharing programs in Europe have been instrumental in encouraging greater EV adoption and making EVs and EVSEs more commonplace. For example, France, the United Kingdom, and Germany all have car-sharing services. Car sharing is expected to remain a popular service in Europe. Forty percent of young adults (aged 18 to 39) in German cities indicate they will be using car sharing services more in the future. Analysts predict car sharing customers in Europe will increase from 1 million to 15 million by 2020 (EVolution, 2014).

Fifty percent of passenger cars in Europe are fleet owned, creating potential for high EV adoption rates. Athlon Car Lease is one of the largest independent car leasing companies in Europe and is supportive of electric vehicle adoption. They created a program called "Fully Charged" to offer the first leasing contracts for EVs. They also formed an agreement with Tesla Motors in

2011 to lease the Tesla Roadster, Roadster Sport, and Tesla Model S to drivers in European countries. Athlon Car Lease and other European fleets have the influence to encourage EV adoption (Tesla Motors and Athlon Car Lease, 2011).

The European Commission and 11 European countries and regions are involved in a program called "Electromobility." This is an initiative aiming to increase the research and development of energy efficient and clean vehicle infrastructure, primarily by providing €20 million for projects (Overview of E+ Partners, 2015). The European Commission has established another organization, the TEN-T Programme, to support the deployment and upgrade of transport infrastructure throughout the European Union, including electric vehicle infrastructure. At the end of 2014, the ELECTRIC project, funded by TEN-T, was announced. This project will build open-access fast charge stations along major highways connecting Sweden, Denmark, Germany, and the Netherlands with 155 charge stations. The five companies collaborating to complete this project are ABB, Fastned, the Swedish utility and e-mobility operator Öresundskraft AB, and the German Testing and Certification Institute VDE Prüf-und Zertifizierungsinstitut GmbH. The project is expected to be completed by the end of 2015 (EU to support, 2015).

11.2.3 Norway

11.2.3.1 Current 2015 Status

A large portion of Norway's electricity production is from hydropower. Norway has an open electric market integrated with other Nordic countries. Norway imports electricity when the price is low (during the night) and exports it when the price is high (during the day) (Market and Operations, 2015).

Norway has the largest per capita fleet of plug-in vehicles in the world with an EV stock of over 40,900 and 6200 public charging stations. Market share sales for EVs were 7.3% in 2013 and 12.5% for 2014 (Global EV Outlook 2015, 2014). The country's fleet is also one of the cleanest in the world since almost 100% of electricity is hydropowered (Countries, 2015). Despite its relatively small size, Norway has been a European leader for EV adoption.

11.2.3.2 Policy

In the Norwegian climate policy, Norway has set carbon dioxide emission targets for passenger cars of 85 g CO_2/km by 2020, 10 g CO_2/km less than the EU target (EV Norway, 2015). The policy outlines incentives to encourage electric vehicle adoption to help achieve the emission target (*Sett 390 S*, 2012).

Norway also has several e-mobility initiatives, including Green Car, ZERO Rally, Taxi Trondheim, and NOBIL. Green Car aims to get 200,000 Norwegians to buy an EV by 2020. They support corporate and municipal fleets with introducing EVs, and work with OEMs and importers to make sure a sufficient number of EVs are in the market. ZERO Rally is an annual rally of

BEVs, PHEVs, hydrogen vehicles, and biofuel vehicles to raise awareness and demonstrate their practicality and user friendliness. Taxi Trondheim was a joint two-year project employing six EVs in a taxi fleet to assess the practical and economic benefits and limitations of EV taxis. NOBIL is a publicly accessible database for charging stations that allows users to build services with free standardized data (EV Norway, 2015).

11.2.3.3 Incentives

A large testament to the EV success in Norway is the incentives the government offers EV drivers, which total to about €17,000. Incentives offered include availability of free public charging stations, toll-free roads, free ferry rides, free parking, and use of the bus lanes. In addition, an import tax is placed on vehicles based on their CO_2 and NO_x emissions, effect, and weight, incentivizing a transition from high polluting vehicles to electric vehicles. Plug-in cars are exempt from paying any vehicle taxes until 2018 as opposed to the high taxes ICE (internal combustion engine) vehicles are charged. In addition, company car taxes are 50% lower for EVs (EV Norway, 2015). A portion of the car taxes is used toward transitioning fleets to clean vehicles (*Sett 390 S* 2012).

In principle, most of these incentives would be just as effective in other countries as they have been in Norway. In practice, though, other countries may have difficulty adopting as similarly pervasive and consistent incentives as Norway.

11.2.3.4 Infrastructure

Ninety-five percent of BEV and PHEV owners have access to home charging and 60% have access to workplace charging. There were 6200 public charging stations total in 2014. CHAdeMO has installed about 80 DC fast chargers (Level 3) and Tesla has installed about 140 Supercharger stalls throughout Norway (CHAdeMO, 2015; Tesla Motors, 2015b). Tesla has plans to install solar power for all Supercharger stations. There do not appear to be many SPCSs in Norway currently, although some stations' electricity use is being offset through solar power generated at offsite locations.

11.2.3.5 Issues and Improvements

The enticing incentives offered in Norway have led to high EV adoption rates that are outpacing EVSE and SPCS infrastructure installation rates. The majority of the population lives in Oslo, the capital, but only 500 public charging stations are located there. More charging stations need to be installed to satisfy the growing number of EVs in Norway.

Despite the positive success of Norway's incentives, some have been coupled with repercussions. Bus lanes in Norway are now mainly comprised of

EVs rather than buses. One day in December 2014, 75% of the vehicles in the capital's bus lanes were EVs while only 7.5% were buses. As the number of EVs on Norway roads increase, this issue will only magnify.

Norway is leading the world in EV adoption and incentives, but it is lacking in charge station infrastructure, especially SPCSs. Although almost 100% of Norway's electricity is hydropower, solar power provides additional benefits beyond generating clean energy. Solar power offsets the load on the electrical grid from charging an electric car. In addition, the solar power could charge EVs during the day without reducing the amount of electricity they export to other countries. If there is excess solar power generated, it could be stored in batteries and used for charging EVs at night to reduce the amount of electricity imported. Also, the stored solar energy could also serve as backup power sources in case of an outage. Solar power has potential in Norway despite the harsh winters. Wind power is another potential energy source to couple with charging stations.

11.2.4 Netherlands

11.2.4.1 Current 2015 Status

The Netherlands has over 43,760 EVs and 12,100 EVSEs. The country aims to have 200,000 EVs by 2020 and 1 million EVs by 2025 (Global EV Outlook 2015, 2014). As of June 2015, the Netherlands comprises about 25% of PHEV sales in Europe (Kane, 2015). Within the country for 2013 and 2014, EVs have maintained 4–5% share of the market sales (Mock and Yang, 2014).

11.2.4.2 Policy

The government launched the National Action Plan for Electric Driving in 2009 to make the Netherlands a world leader for electric driving. The plan has the government spending about €65 million in support of this effort. In order to further encourage EV adoption, the Netherlands has the highest tax rate on petrol of any EU state. In fact, it is double the rate of Bulgaria, the EU's lowest state tax rate on petrol (Heymann, 2014). More broadly, the Netherlands has a goal of cutting GHG (greenhouse gas) emissions 25% below 1990 emission levels by 2020, and encouraging EV adoption will help achieve this (Climate Case, 2015).

11.2.4.3 Incentives

The Netherlands government and many cities offer incentives to encourage EV adoption. Full electric lease cars have less income tax addition ranging from 0–14% compared to 14–25% for ICE cars in 2013 to 2015. However, this tax break will end in 2016 and the income tax addition for PHEVs will increase to 14–21%. Amsterdam offers additional incentives including subsidies and benefits. The city provides a subsidy on EV purchase prices of €5000 for passenger cars, €10,000 for taxis, and €40,000 for trucks. Amsterdam also allows EVs to have no waiting list for parking permits, to charge for free in certain parking garages, and to pay no registration or annual circulation tax (EVolution, 2014).

In addition, the Dutch Consortium for the Tender of Electric Cars (DC-TEC) worked in 2011 to provide assistance to participating companies buying at least 10 EVs. The goal was to deliver at least 3000 EVs at a price comparable to ICE vehicles to encourage initial fleet EV adoption (Countries, 2015).

11.2.4.4 Infrastructure

In 2009, Netherlands developed the Open Charge Point Protocol (OCPP), a protocol accepted in 50 countries and over 10,000 charging stations. The goal of the OCPP is to ensure flexibility of EVs between different charging networks worldwide, providing accessibility, compliance, and uniform communications between charge stations and management systems. The Open Charge Alliance is an industry alliance comprised of EV charging hardware, software vendors, and charging network operators and providers, who are determined to uphold and foster global development of the OCPP (Open Charge Alliance, 2015).

ElaadNL and EVnetNL work together to create a network of over 3000 public charging stations for EVs in the Netherlands. These initiatives coordinate connections for public charging points and enable the collaboration of different Netherland network operators and managers (ElaadNL, 2015). The E-laad Foundation is a consortium initiated in 2010 supported by cooperating regional grid operators to increase EV infrastructure. The objective is to develop 10,000 public charging points. Municipalities can request up to 2000 charging spots, and EV drivers can request 8000 charging locations. The cooperating grid managers fund these charge station installations with a €25 million budget. EV drivers can request to have an EV charge station installed at their home, workplace, or a different location and the installation will be done for free in a reasonable amount of time (Countries, 2015).

One specific successful charge station company in the Netherlands is Fastned. Fastned fast chargers are Level 3 charge stations installed with a solar canopy to offset electricity demand. See Figure 11.1. The company plans to build about 200 stations along major highways, aiming to have about half completed by the end of 2015. It is financially sustained through investments and customer pricing plans. Fastned has received about €15 million in investments using several incentives. First, Fastned offers certificates of €10 apiece as one share of the company and allows any certificate holder to attend certificate holder meetings and to vote on FAST board members. This system allows almost anyone to purchase a share and voice their opinion if so inclined. Second, Fastned offers a deal for shareholders who

FIGURE 11.1
The Fastned charge stations are a growing charging network powered with renewable energy. *Photo credit*: Fastned fast charging station Vundelaar, Roos Korthals Altes.

purchase over 2500 certificates to become members of the Fastned Founders Club, which allows members lifetime free charging. The perks of this membership encourage greater company investment, which fund their charge station development. The pricing plans, which also fund Fastned, allow consumers to choose either to pay a price per kWh or purchase a subscription plan. Fastned expects to reach the breakeven point when there are about 50,000 EVs using their network. Their business model resembles the Tesla Superchargers model, in that they are focused on Level 3 charging, solar power, and building up the charge station network (Fastned, 2015). There are only a few Tesla Superchargers (5 as of July 2015) distributed throughout the country, which will be solar powered in the future (Tesla Motors, 2015b).

In June 2015, Stedin, in cooperation with the city of Utrecht and other partners, installed the first vehicle to grid (V2G) capable solar powered charging station in the Netherlands (World Premiere Utretcht, 2015). This will allow solar energy to be stored in EV's batteries and delivered to the grid if needed.

The Netherlands has made vast developments in the charge station infrastructure. The country is building an infrastructure with many Level 3 stations, locating them at advantageous points, and coupling a large number of the stations with solar energy. Netherlands has also started to construct charge stations with V2G technology *and* solar power (Amsterdam Electric, 2015). The country will hopefully continue this progressive path to satisfy the increasing demand of EVs and prepare for the future.

11.2.5 United Kingdom

11.2.5.1 Current Status

The United Kingdom has 21,430 EVs and 2870 EVSEs (Global EV Outlook 2015, 2014). In Western Europe, 40.6% of all PHEVs sold in the first four months of 2015 were sold to people in Britain. The UK and the Netherlands together comprise over 67% of EU PHEV sales (Kane, 2015).

The UK primarily produces electricity using fossil fuels; coal and natural gas. About 15% of electricity is generated using nuclear energy and 7% generated using renewable energy sources. The UK electricity system is connected to France and Ireland, permitting electricity to be imported and exported when it is most economical (Electricity Generation, 2015).

11.2.5.2 Policy

The UK aims to reduce GHG emissions by at least 80% below 1990 emissions by 2050. In 2011 the Carbon Plan was created, which included several transportation actions. One action included the provision of over £400 million in funding between 2011 and 2015 for EV recharging infrastructure, research, and development to support ultra-low emission vehicles. The Carbon Plan also included the electric train routes and funding for low carbon emission buses and heavy goods vehicles. The government has an additional goal for all cars and vans to be zero emission vehicles by 2050 (2010 to 2015 government policy, 2015).

11.2.5.3 Incentives

In order to support EV sales, the UK government provides a one-time premium of £4000–7000, depending on vehicle price, to car owners whose vehicles emit less than 75 g CO_2/km. The government offers a Plug-in Car Grant and Plug-in Van Grant, which cover 35% of the EV car purchase price (maximum of £5000) and 20% of the EV van purchase price (maximum of £8000), respectively. Also, the city of London exempts EV drivers from paying the congestion charge and road tax when driving in the Congestion Charge Zone (EVolution, 2014).

11.2.5.4 Infrastructure

The UK has over 57,500 public EV charge stations. As UK EV sales continue to increase, charge station installation rates must follow as well. The UK also has several initiatives to support charge station infrastructure. For example, Plugged-In Places (PIP) was designed to spur charge station development in several hubs. The government provided millions to this program to fund the construction of thousands of charge points nationwide and to provide free charging. Once the initiative ended in

2014, charge station companies started charging fees for charge station use. The Committee on Climate Change, which advises the government, has projected that a charging infrastructure could be supported for almost 2 million EVs in the UK by 2020 at a cost of a few hundred million pounds (Countries, 2015). The government is focused on building home charging infrastructure and workingplace charging infrastructure. The National Planning Policy Framework encourages local authorities to consider implementing policies that include charge station infrastructure in new domestic developments. In addition, EV charge stations were granted permitted development, which is automatic planning permission for landowners to install charge points, enabling simpler and quicker installation processes (Making the connection, 2011).

Ecotricity is a utility company in the UK that solely uses clean energy, primarily wind energy but a portion of solar, tidal energy, and biogas as well. They are a not-for-dividend company meaning they have no shareholders. Instead, they are funded by customers' energy bills. The utility powers and heats homes and businesses, and is also building an EV charge station network, powered by clean energy, along highways in the UK. There are over 100 charge stations in the Ecotricity network. Other networks' swipe cards work for these charge stations too. Currently the charge stations are free to use for all customers, although a fee may be imposed later in the future (Ecotricity, 2015).

Green Motion is a car and van rental company, which has EVs, PHEVs, and hybrids in their fleet. The charge stations use solar or wind power if appropriate (Green Motion, 2015). In addition, there are several Tesla Superchargers distributed throughout the country (Tesla Motors, 2015b).

11.2.5.5 Issues and Improvements

London has had difficulty maintaining functionality of the city's charge stations. Beginning in 2014 and continuing through 2015, 20–30% of London's public charge stations have consistently been out of service for maintenance at any given time. On-street charging is the primary mode for EV owners to charge their vehicles, given the city's limited garage or driveway parking. Bluepoint, a private operator who recently bought Source London, is responsible for repairing the network, but is continuing to struggle to do so. The poor charge station network maintenance affects EV drivers, EV sales, and companies such as rental companies planning to add EVs to their fleets (Sharman, 2015). Bluepoint needs to adopt a better maintenance repair system for London's electric charge stations in order to support EV industry growth.

In addition, the UK does not have many charge stations powered with solar energy or other renewable energy forms. A greater number of SPCSs should be constructed and located throughout the country. If not, the EVs are charged mainly using fossil fuels, which still release carbon emissions.

11.2.6 France

11.2.6.1 Current Status

The government aims to have 2 million EVs by 2020. Several car manufacturers and fleet services are leading the initiative, and are pledging to produce or purchase mass orders of EVs (Countries, 2015).

La Poste is a postal delivery service company in France with the nation's largest corporate car fleet and the world's largest electric vehicle fleet. The company currently has 5000 EVs with plans to deploy 10,000 EVs in total. La Poste is also testing extended-range hydrogen electric vehicles in 2015 and assessing their viability (Le Groupe La Poste, 2015). This company has played a large role increasing EV adoption in France.

11.2.6.2 Policy

France has a 14-point plan to encourage hybrid and electric vehicle adoption. Plan elements include building charge station infrastructure in 2010 with a €70 million budget, Renault establishing a lithium ion battery factory in Flins to produce over 100,000 batteries annually, requiring offices and homes built in 2012 or later to have charge stations integrated, and recycling used batteries from electric vehicles (Detailed presentation, 2012).

France reduced its GHG emissions by 13% from 1990 to 2012; however, emissions in the transportation sector were not reduced during this time. In July 2015, France passed the Energy Transition Law, which aims to reduce France's environmental impact, especially concerning energy and emissions. The law calls for a 40% reduction of GHG emissions between 1990 and 2030 and to further reduce emissions by 25% by 2050. It quadruples France's fossil fuel carbon tax by 2020. In addition, it calls for the country to cut 50% of its energy usage by 2050 and increase the energy mix's share of renewables to 32%. The country's energy mix as of 2015 is primarily nuclear energy with smaller portions of renewable and fossil fuel energy generation as well. The Energy Transition Law also works to develop France's clean transportation and energy efficient mobility. It states that fleets and their public institutions should have 50% of their vehicles be clean vehicles and announces that the country will deploy 7 million electric vehicle charging stations. Since 2010, the government has already been practicing some infrastructure provisions such as enabling local governments to install public charge stations and obliging them to do so at public parking areas, reserving a quota of parking spots for EVs and charge stations at workplaces and shopping centers, and requiring builders to install charge stations on inhabitants' requests (Energy: energy, 2015).

11.2.6.3 Incentives

Vehicles that emit 20 g CO_2/km or less receive a €6300 premium, vehicles that emit between 20 and 60 g CO_2/km receive a €4000 premium, and vehicles that emit between 61 and 110 g CO_2/km receive a €2000 premium. In addition, hybrid vehicles, which emit less than 110 g CO_2/km, are exempt from company car tax during the first two years after registration. All electric vehicles are totally exempt from the tax (Overview for purchase and tax incentives, 2015). Furthermore, the government of France is offering a bonus of €3700 *on top of* the premium to drivers who trade a diesel car that is 13 years or older for an electric vehicle and a bonus of €2500 to drivers who trade for a plug-in hybrid (Ayre, 2015a). On top of these national incentives, drivers who live in the Haute-Normandie region and purchase EVs also receive €5000 for individuals and up to €25,000 for companies and communities. Schools in this region that purchase EVs will also receive support for 70% of the vehicle cost and total cost support for charge station infrastructure (Gordon-Bloomfield, 2014).

11.2.6.4 Infrastructure

The government would like to have 7 million charge stations installed by 2030. In order to get there, France developed a strategic roadmap in 2009 for the expansion of the charge station infrastructure. Three key parameters were identified: creating standards for charge stations to ensure interoperability and flexibility, developing long-term, viable economic and business models for vehicles and infrastructure, and matching infrastructure supply with demand. The country recognized the importance of the government's support of the technology in the beginning development stages, until 2020. After that, leading up to 2050, viable business models would have to start solely supporting the industry (Strategic roadmap, 2015).

Autolib' has played a large role in building France's charge station infrastructure and increasing EV awareness. Autolib', created by the city of Paris and Bolloré, is an electric car sharing service in Paris that allows customers to pay a subscription fee to drive an electric Bluecar throughout the city and park at designated spots. Each parking spot has a charge station. There are over 2200 Bluecars and 4300 charge stations now deployed in Paris. Certain other EVs are capable of using the Autolib' charge stations as well, including the Nissan Leaf and Mitsubishi I-Miev. Bolloré has created similar car sharing services in Lyon, France and soon in London (Autolib', 2015). The company has helped build a portion of France's (and soon London's) charge station infrastructure, increased awareness of EVs with high visibility, and enabled thousands of customers to experience operating an EV. Autolib' has normalized electric vehicles in Paris. Yet another group, AdvanSolar, is a startup company that will install solar powered charge stations in France. Currently, there is one AdvanSolar station installed in the city of Nice (Advansolar, 2015).

Eco2Charge is a growing project in France, coordinated by Bouygues Energies & Services and eight partners, with an aim to increase EV charge station infrastructure. This project focuses on reusing EV batteries, after their primary use in a vehicle, as local energy storage systems for charge stations. The batteries will store electricity during the night, when electricity prices are lower, and then charge vehicles with the stored energy during the day. These charge stations will be primarily located at work places, parking lots, campuses, and other locations where fleets park. Installations will begin by the end of 2016. Recycling used EV batteries is a cost-effective storage method. In principle, this project could be further improved if solar panels were added to the charging stations. Solar power would charge plugged in vehicles, in addition to the batteries, during the day, reducing reliance on the batteries and enabling a greater number of EVs to charge. Wider future applications of the Eco2Charge program include utility-scale storage systems where large solar or wind power stations are coupled with 50+ recycled EV batteries. The system could then use stored renewable energy to act according to grid-wide demand, such as adding electricity to the grid at peak times (Eco2charge, 2015). Last, Tesla also has a presence in France, with several Superchargers installed (Tesla Motors, 2015b).

11.2.6.5 Issues and Improvements

There are few solar powered charge stations developed in France. Although a couple of companies are working to deploy more SPCSs, there needs to be greater support. Powering more charge stations with solar power will reduce the strain on the electrical grid and help meet France's Energy Transition Law requirement of increasing the share of renewable energy in the energy mix to 32%.

11.2.7 Germany

11.2.7.1 Current Status

Germany aims to have 1 million EVs on the road by 2020. The country had 24,420 EVs and 2820 EVSEs at the end of 2014 (Global EV Outlook 2015, 2014).

Fossil fuels, renewable energy, and nuclear power are the main power sources for the country's electricity production. Germany plans to phase out nuclear power.

11.2.7.2 Policy

Germany aims to cut 40% of GHG emissions by 2020 and at least 80% by 2050 compared to 1990 levels (Energy concept, 2015). Transitioning to electric vehicles charged with renewable energy is the country's main strategy to reaching these targets. Thus far, Germany has been reaching this goal. Germany, along with France and Italy, has some of the largest reductions of carbon emission from newly registered passenger vehicles for the European Union (EVolution, 2014). Every German auto manufacturer has a hybrid electric vehicle on the market and also offers or is developing an all-electric vehicle (Countries, 2015).

Four ministries are involved in Germany's electromobility: the Federal Ministry of Economics and Technology (BMWi); the Federal Ministry of Transport, Building, and Urban Development (BMVBS); the Federal Ministry for the Environment, Nature Conservation and Nuclear Safety (BMU); and the Federal Ministry of Education and Research (BMBF) (Countries, 2015).

Germany has established the National Electromobility Development Plan to increase the country's electromobility and help achieve the goal of 1 million EVs by 2020. The plan focuses on funding research and development and practicing different market strategies to facilitate electric vehicle adoption. The National Electric Mobility Platform brings scientists, politicians, and industry workers together to achieve the National Electromobility Development Plan objectives. Germany aims to become a leading market and provider in the electric mobility sector (National Electromobility Development Plan, 2015).

There are three stages to Germany's goal of having 1 million EVs by 2020. These stages are research and development (2014), market expansion with vehicle and infrastructure policies (2017), and launching a mass market (2020) (Vergis et al., 2014).

11.2.7.3 Incentives

The government included €500 million for the development and commercialization of electric vehicles and infrastructure in its Second Economic Stimulus Package. The government considered not offering consumers incentives for purchasing EVs, but eventually decided to provide a tax incentive to spur greater EV adoption. All BEV vehicles licensed before December 31, 2015 are exempt from taxes for 10 years and those licensed between January 1, 2016 to December 31, 2020 are exempt for five years (Block, 2015; Countries, 2015).

The Electromobility Act was passed in early 2015 providing municipalities the right to grant special perks for EV users. The benefits include reserved and often free parking, use of bus lanes, and special transit passes for pollution sensitive areas (Tost, 2014a).

11.2.7.4 Infrastructure

A program of the National Electromobility Development Plan is Electric Mobility in Pilot Regions, which focuses on building the charge station infrastructure in Germany. The government has allocated €130 million to this program for the eight regions with pilot infrastructure projects. The program has installed over 2000 charge stations. Car sharing services have also built charge station infrastructure. Flinkster car-sharing service is available in over 140 towns and cities in Germany. There are over 500 electric vehicles in their fleet and over 800 charge stations in their network (EVolution, 2014).

In order to increase the charge station infrastructure, the government has enlisted a highway service provider, Autobahn Tank & Rast GmbH, to build 400 fast-charging station sites along highways by the end of 2017 (Cremer et al., 2014).

European Utility RWE has helped develop the largest public charging network in Europe. RWE and other major German utilities are working with auto manufacturers and companies to increase charge station infrastructure. RWE is working with Daimler in a joint venture called E-mobility Berlin to install 500 intelligent charge stations powered with renewable energy. These charge stations allow consumers to add the cost of EV charging directly to their utility bill in addition to monitoring electricity rates and battery status (RWE-Mobility, 2015). Hubject is another joint project with car manufacturers BMW and Daimler, utilities EnBW and RWE, and companies Bosch and Siemens. Hubject invited Open InterCharge Protocol (OICP), which allows information to be exchanged between an EV and its driver, a charge station, and a utility. Hubject network is primarily in Germany but has a presence in neighboring countries such as Belgium, Austria, and Finland as well (Masson, 2013).

The Berlin based company Younicos AG has a solar powered charging station called Yana. This station has a solar array that tracks the sun and a 100 kWh battery to store solar energy. The station can charge up to eight vehicles at once. Younicos displays advertisements on the charge stations to generate additional revenue for the installer (Solar charging station, 2010). Finally, Tesla plans its Supercharger network to have complete, sufficient coverage of the country by the end of 2015 with 40–50 Superchargers installed total (Tesla Motors, 2015b).

11.2.7.5 Issues and Improvements

Germany was planning to have 100,000 EVs on the road by 2014, but only reached 24,420 EVs. Lack of incentives and infrastructure were two reasons for low EV adoption rates. The National Electric Mobility Platform (NPE) commented on Germany's progress at the end of 2014 and suggested steps to achieve their goal of being a leading manufacturer and adopter in electric transportation (Tost, 2014b). Suggested recommendations include incentivizing adopters with bonus tax depreciations for commercial users and implementing the Electromobility Act, building up the public charge station infrastructure, initiating public and private procurement of EVs, and manufacturing batteries in Germany. Germany adopted the Electromobility Act in early 2015 and will monitor the affect on EV sales. The additional NPE recommendations should be adopted as well, especially building the public infrastructure. Germany should also focus on coupling the charging infrastructure with solar panels, since the country's main strategy to decrease GHG emissions is increasing electromobility and renewable energy (Tost, 2014a).

11.2.8 Denmark

11.2.8.1 Current Status

Denmark aims to have energy independence in the future from fossil fuels. Denmark plans to couple wind power with EVs and V2G technology to meet electricity and transportation goals. As of December 2014, Denmark has 2800 EVs and 1720 EVSEs (Countries, 2015).

11.2.8.2 Policy

In the Danish Energy Agreement, Denmark aims to reduce carbon dioxide emissions to 34% lower than 1990 levels by 2020. Denmark also has a goal to increase the energy share of renewable energy to 35% by 2020. Wind currently accounts for over 20% of electricity production, but the Danish Energy Agreement includes a goal to increase this percentage to 50% by 2025. The Agreement also identifies the future need of a smart grid with increased renewable energy production and EVs; thus, it calls for the development of a smart grid strategy and strategy for energy-efficient vehicle promotion. €9.4 million is allocated for EV, hydrogen, and gas infrastructure (Addressing climate change, 2015).

In 2008, Danish Parliament entered a Climate and Energy Agreement, leading to the Danish EV promotion program. As part of the agreement, €4 million was allocated to EV battery demonstration programs, administered by the Danish Energy Agency. The program is designed to hear user feedback about EVs to become aware of deployment barriers. The following year the Danish Transport Authority created the Centre for Green Transport to manage sustainable transport initiatives. The Centre conducts demonstration projects to promote environmentally aware and energy efficient transportation solutions and test projects with EVs and alternative fuels. Examples of these projects include testing hybrid buses and leasing EVs on a timeshare basis. The EDISON project administered by Energinet.dk, the Danish transmission system operator, also began in 2009 to develop system solutions and technologies for EVs and PHEVs. The project connects research institutions with industry to conduct research, development, and demonstration. However, the majority of research in Denmark has been placed on hybrids and hydrogen-fuel cells.

The Danish Electric Vehicle Alliance is an alliance of companies across the EV value chain including energy companies, charge station infrastructure operators, and EV manufacturers (Countries, 2015). The group works to foster the growth of EVs in Denmark while representing the interests of member companies (Mission, 2015).

11.2.8.3 Incentives

Incentives include 25% VAT (value-added tax) for vehicles weighing less than 2000 kg and exemption from the registration fee and annual circulation tax based on fuel consumption. Total incentives for the BEV are about €15,500 (private) or €3300 (company) and for the PHEV are about €22,800 (private) or €3400 (company). Unlike most countries, Denmark's incentives are higher for private users rather than companies. This different incentive strategy does not seem to work in Denmark's favor, however, since EV adoption is low compared to other countries (Mock and Yang, 2014).

11.2.8.4 Infrastructure

CLEVER is a Danish company that established Denmark's first nationwide fast charging network. These stations are installed in convenient locations such as highways, shopping centers, restaurants, workplaces, and homes. CLEVER worked on Test-an-EV, Northern Europe's largest research project, supported by the Danish Energy Agency. The project had 200 EVs, which 1580 participants drove and reported their experience on for 3 months. The majority of charging occurred at home. Seventy-one percent of the project participants reported a positive opinion of electric vehicles. Users felt that the vehicles' range met their daily driving needs. CLEVER also has partnered with several different entities to build the charge station network, including Volkswagen auto manufacturer, Shell oil company, and the Swedish utility Öresundskraft (CLEVER, 2015).

E.ON is another company building the Denmark charge station infrastructure with 700 charge stations as of 2015. The electricity used to charge these vehicles is renewable since it is generated using hydropower. E.ON installs stations in the city, homes, and for businesses (About E.ON, 2015). Tesla also has several Superchargers installed in Denmark (Tesla Motors, 2015b). However, CLEVER and E.ON are the main builders of the Denmark infrastructure.

11.2.8.5 Issues and Improvements

Denmark has a low EV and EVSE stock. The country has a relatively similar population size as Norway, but had 38,000 less EVs and 4500 fewer EVSEs at the end of 2014 (Global EV Outlook 2015, 2014). In order to keep up with other countries, Denmark must increase charge station infrastructure, especially SPCSs, and encourage much greater EV adoption rates.

11.3 Asia

The Electric Vehicle Association of Asia Pacific (EVAAP) is an international membership organization that works to promote the adoption and development of EVs and HEVs in Asia and the Pacific region. EVAAP exchanges relevant information among members, collaborates with other international groups with similar goals, and educates the government and public. Members include China, Japan, Hong Kong, and Korea (EVAAP, 2015).

Asian countries have had high adoption rates of e-bikes and e-scooters. These vehicles are fuel efficient, easy to maneuver in congested traffic, and more affordable than a car. China is the largest market for e-bikes and e-scooters and India is the second largest. E-bikes are the most popular mass-produced alternative fuel vehicle in China. Over 20 million e-bikes are sold each year (Bae and Hurst, 2012). Although they can travel only 20–30 miles per hour, they are much more affordable, starting at $1000, than EVs (Timmons, 2013). However, in most other Asian countries, gasoline two-wheel vehicles are still preferred over electric (Bae and Hurst, 2012).

Despite the high adoption rates for e-bikes and e-scooters, passenger vehicles are still widely used in Asian countries. It is important to consider the progress of EVs and improvements that can be implemented to encourage greater EV adoption.

11.3.1 China

11.3.1.1 Top EV and PHEV Car Sales

The top three electric vehicles sold in China as of June 2015 are BYD Qin PHEV, BAIC E-Series EV, and Zotye Cloud (EV Sales, 2015). BYD Qin, a domestically manufactured vehicle, has been dominating the Chinese EV market since 2014. BYD Qin is comparable in size to the Mitsubishi Outlander PHEV. The automaker BYD is rivaling Tesla. The BYD e6 is the second best-selling vehicle in China and the 2016 model will have a 250-mile range, similar to Tesla's Model S 270-mile range, but $20,000 cheaper. In addition, the company is building a battery factory that could rival Tesla's Gigafactory (DeMorro, 2015b; Shahan, 2014).

In 2014, five domestic EV manufacturers BYD, Kandi, Chery, Zotye, and BAIC held the top five highest sales in China. Overall, Chinese-manufactured EVs seem to be the leading vehicles sold in China, outselling Tesla and other foreign manufacturers (Tillemann, 2015).

11.3.1.2 Current Status

At the end of 2014, China had 83,200 EVs and 30,000 EVSEs, comprising 12% of the global 2014 EV stock and ranking third highest EV stock in the world (Global EV Outlook 2015, 2014). China also had 36,500 EV buses and 230 million electric bikes at the end of 2014 (Global EV Outlook 2015, 2014). China is projected to have over 350 million electric two-wheel vehicles sold by 2018 (Bae and Hurst, 2012). There was a boom in the electric vehicle market in China in 2014, which *The China Electric Vehicle Charging Station and Charging Pile Report 2015–2016* attributes to Tesla (China electric vehicle, 2015). Electric vehicles and PHEVs are expected to continue to be subsidized relatively equally despite PHEVs' lower total operating cost (Gao et al., 2015; Krieger et al., 2012).

China's primary electricity source is coal, followed by hydropower, natural gas, nuclear and other renewable sources. Twenty-six percent of electricity produced in China is renewable (China, 2015).

11.3.1.3 Policy

China targets carbon dioxide emissions to peak by 2030 or earlier and to decrease the carbon intensity of gross domestic product (GDP) by 60–65% below 2005 levels by 2030 (Climate Action Tracker, 2015). The State Council's Notice on Energy Conservation and New Energy Vehicle Industry Development (2012–2020) called for the production and sale of 500,000 EVs and HEVs by 2015 and 2 million by 2020. China failed to meet the 2015 target and does not seem on track for 2020. Only existing Chinese auto manufacturers were eligible to produce electric vehicles until early 2015 when the National Development and Reform Commission (NDRC) modified rules to allow all manufacturers with special licenses to produce EVs (Marro et al., 2015).

The National Government Offices Administration declared that at least 30% of government cars purchased annually must be BEVs, PHEVs, or other new non-polluting vehicles produced in China. In addition, these government offices are required to install charge station infrastructure for their EVs and improve other clean vehicle infrastructure as well (China urges, 2014).

Four ministries of the central government issued the Interim Measures for Financial Aid Fund Management of Electric Vehicles Private Purchase in Pilot Cities (2010–2012). This provided a subsidy to EVs for private purchase or use in the pilot cities, a grant allowance for power battery production, battery charging stations, and infrastructure standardized construction, and funds for directory review and inspection (Tan, 2014).

11.3.1.4 Incentives

Consumers who purchase a locally-produced new-energy vehicle (NEV), BEVs and PHEVs, receive a ¥35,000–60,000 ($5650–9700) national subsidy until 2020. After 2020, this national subsidy will be scaled down each year. In addition, domestic made NEVs will be exempt from the purchase tax until 2017. Local governments are providing additional incentives as well. For example, the Shanghai government offered free license plates for new Tesla vehicles, which saves the driver ¥74,000 ($12,000). Other local governments are choosing to match the national subsidy. Some local governments are also choosing to cap new vehicle registrations and reserve a portion for EVs (China urges, 2015).

In Beijing, odd number license plates and even number license plates must alternate the days they drive on freeways. However, Beijing is adding an additional incentive for EVs by allowing EV drivers to use freeways every day (DeMorro, 2015a).

The majority of incentives in China only apply to locally manufactured EVs, not foreign vehicles. The lack of incentives and high import duties are hurting some foreign EV sales such as Tesla (Tillemann, 2015).

11.3.1.5 Electric Transportation

Auto manufacturer BYD sells 4000 electric buses in China and several thousand more worldwide annually (Loveday, 2015b). Nanjing Public Transportation Group Co., LTD has purchased some of the largest electric bus and taxi orders from BYD. The company aims for its fleet to become all-electric and one of the largest in the world (Nanjing Public Transportation Group 2015). Also, the city of Wuhan has added thousands of BYD electric taxis to their fleets (Morris, 2015).

11.3.1.6 Infrastructure

China has focused on building the EV charge infrastructure using charge stations and battery swapping. The State Grid Corporation of China and China Southern Power Grid, the two main operators of China's power grid, have dominated building China's charge station infrastructure. These state-owned companies have each focused on different regions to build charging infrastructure. Both have installed charging stations, charging and battery swap stations, and developed smart charging and swap networks (China electric vehicle, 2015). Other key players in the charge station infrastructure development include several electric companies, technology companies focused on smart grid and renewable energy distribution, and a public transportation company. In 2014, the State Grid opened the distributed power grid and charge station market to private investors to stimulate installation rates. This allows consumers to install a charging station or residential solar power, and enables privately owned small-scale power sources to integrate into the state grid (China electric vehicle, 2015).

While the majority of EVs in China seem to use charge stations, some EVs and most electric buses use battery swapping (Loveday, 2015a). In April 2015, Ziv-Av Engineering (ZAE), an Israeli engineering firm, signed a deal with China's Nanjing Bustil Technology, an electric vehicle focused company, to create 7000 battery swap stations for electric buses in Nanjing, China. If successful, the Chinese company may expand to electric taxis as well (ZAE, 2015).

In 2014, Tesla partnered with Hanergy, a Chinese clean energy and thin-film solar power company, to build the first solar powered Superchargers in China (Tesla Motors, 2015a). There are also nonsolar powered Superchargers in China (Tesla Motors, 2015b).

As a result of all the above actions, the number of charging stations for electric two-wheel vehicles jumped between 2010 and 2014 from 76 to 723 charging stations (China electric vehicle, 2015).

11.3.1.7 Issues and Improvements

The Chinese government seems to be focused on the luxury, higher-end EVs, providing generous subsidies for these vehicles. A majority of Chinese citizens cannot afford these high-end EVs and are unable to purchase an EV. Greater support must be given to the average consumer and more affordable EVs to help increase consumer EV purchases. If the government aims to build an EV fleet suited for its domestic needs, it should also improve international vehicle sales and increase small-scale EV production (Akcayoz De Neve, 2015).

Given the popularity of e-bikes and e-scooters in China, the bike charging infrastructure should also be expanded. In 2014, there were over 700 charging stations in China, but given projections for almost 60 million two-wheel electric vehicle sales in 2016, the charging infrastructure for these vehicles must increase rapidly. China is focusing on expanding EV sales, but two-wheel vehicles may potentially be the future for the country. An equal or greater focus should be placed on these electric two-wheel vehicle sales and infrastructure.

Electricity in China is primarily produced from coal. Although vehicle emissions are reduced when transitioning to EVs, air pollution problems remain when the EVs are recharged using coal power, and pollution will possibly increase with the greater electricity demand. In order to avoid this, China should increase the share of renewable energy in the energy mix. In addition, China should couple solar power with charging and battery swapping stations to avoid an increased demand on the grid and recharge EVs using clean energy. Solar power would also reduce the demand surge on the grid, avoiding voltage fluctuation and poor voltage quality (SGCC Decontrols, 2014).

Range anxiety is still a large limitation to consumer EV adoption in China. There is insufficient charge station infrastructure. Often an area has limited charge stations built or the EV is not compatible with the charge station. Cities have led the effort of building infrastructure, which has led to varying standards city by city and local protectionism. For example, auto-manufacturer BYD needs approval from every city EV promotion office before entering a conversation of building infrastructure. China is working to reduce local protectionism and establish a national charging standard for EVs. In 2014, China announced it would unify its charging standards with Germany to the 7-pin Type 2 plug also used in the Netherlands. Another obstacle is that it is difficult to secure physical locations in China's cities since parking spaces and gas stations are rare. Earlier in 2014, 1400 people in Beijing were granted free EV licenses, but 70% gave them up because there were no realistic places to charge in the city. China should continue building its national charge station infrastructure, using one charging standard, to encourage greater EV consumer adoption and to alleviate range anxiety (Yu, 2015). The next challenge to tackle should be finding or creating suitable charging locations in cities (China Top Sector E-mobility, 2014).

11.3.2 Japan*

11.3.2.1 Top EV Car Sales

As of June 2015, the top EV car sales for Japan are the Nissan Leaf, Mitsubishi Outlander PHEV, and Toyota Prius Plug-in (EV Sales, 2015).

11.3.2.2 Current Status

Japan is ranked second in the world in terms of EV stock size for 2014. Japan has 16% of the 2014 EV global stock with 108,250 EVs and 11,500 EVSEs.

Japan has few domestic energy resources; it is the third largest net oil importer in the world behind the United States and China. Japan is the largest importer of liquefied natural gas and the second largest importer of coal. Nuclear power was a main electricity source in Japan until the Fukushima plant accident. Imported natural gas, crude oil, fuel oil, and coal make up for the nuclear power loss, resulting in higher consumer electricity prices and a greater reliance on fossil fuels. Japan has plans to use nuclear energy as a baseload power source and to balance the energy mix with more renewable and alternative energy sources (Japan, 2015).

11.3.2.3 Policy

In July 2015, Japan took a step backward by reducing the country's emissions reduction target from 25% below 1990 levels by 2020 to 26% below 2013 emission levels by 2030 (equivalent to 18% below 1990 levels by 2030). This target is easier to reach and can be met using the current policies in place without taking further action (Climate Action Tracker, 2015).

In 2006, Japan set a target to focus on battery research and development to improve the lithium-ion battery and to promote EV sales to have economy of mass production for the batteries. In 2010, Japan developed the Next-Generation Vehicle Strategy. This Strategy aims for Japan to have 20–50% of the country's new vehicles as clean vehicles (hybrid, EV, fuel-cell, clean diesel) with 15–20% BEVs and PHEVs by 2020. In addition, by 2030 it aims for 50–70% of new vehicles to be clean vehicles, with 20–30% BEVs and PHEVs.

* Japan defines Next Generation Clean Energy Vehicles as HEVs, PHEVs, BEVs, natural gas vehicles, clean diesel vehicles, and fuel cell vehicles.

The strategy also includes goals to install 2 million charge stations and 5000 fast charge stations in Japan by 2020 (Government initiative, 2015).

11.3.2.4 Incentives

In fiscal year 2014, the government offered purchase subsidies of ¥850,000 for PHEVs and BEVs and ¥350,000 for clean diesel vehicles. The incentives for the acquisition tax apply from April 2012–March 31, 2015 and for the tonnage tax apply from May 2012–April 30, 2015. Next generation vehicles are exempt from the acquisition tax and exempt from the tonnage tax for the first two vehicle inspections.

The government allocated ¥100.5 billion in fiscal year 2012 to subsidize charging station installations from March 2013–December 2015. Charge stations installed in accordance with a governmental or highway operating plan had 75% of the cost subsidized and charge stations installed for public use, parking lots, or other uses had 50% of the cost subsidized. Some cities are providing additional subsidies to encourage charge station installations. For example, Tokyo is offering financial incentives to businesses that install charge stations (Nagatsuka, 2014).

11.3.2.5 Infrastructure

Japan has more EV charging points than gas stations. Japan has 40,000 EV charging points, including those installed at home (although not electrical outlets), and 34,000 gas stations. This impressive statistic indicates Japan's dedication to increase EV charge station infrastructure and transition to electric vehicles (Ayre, 2015b).

A major strategy for building the charge station infrastructure is the EV and PHEV Town Concept. Fifteen towns were selected for projects to demonstrate EV and PHEV deployment and are used as models for nationwide charge station infrastructure. Several towns with high tourism promote electric rental cars and electric taxis and have fast charging charge stations located in popular tourist spots. In addition, several cities have EV buses and EV taxis also. The infrastructure is slightly customized for each town depending on people's daily needs (Our best practices, 2015).

CHAdeMO is a Japanese company that sells fast chargers internationally. The charge stations allow bidirectional charging and are Vehicle-to-Home (V2H) capable. Over 5400 CHAdeMO fast chargers have been installed in Japan (CHAdeMO, 2015).

In 2014, several auto manufacturers, Toyota, Nissan, Honda, and Mitsubishi, created a joint company called Nippon Charge Service. This company is dedicated to increasing Japan's charge station infrastructure by funding a portion of applicants' installation costs. These charge stations are then a part of the Nippon Charge Service network. The company will focus on building infrastructure in highly trafficked areas such as fueling and service stations, along highways, parking lots, hotels, and larger commercial facilities (Japan automakers advance, 2014).

Tesla is also installing Superchargers in Japan and powering them directly or indirectly with solar power to have no impact on the grid (Tesla to invest, 2015).

Toshiba has been testing solar powered charging stations for ultra small EVs in Miyakojima City, an island in Japan. Given the frequency of power outages on the island, Toshiba plans to use storage batteries as emergency power sources. Data will be analyzed from these test projects for future development on the island (Toshiba Corporation, 2013).

KYOCERA Solar Modules Company has provided solar panels for several EV charging stations worldwide. In Japan, KYOCERA worked with manufacturer Shintec Hozumi to charge EVs parked at the company's headquarters using solar power. Solar panels of 230 kW generation capacity were installed on the building rooftop and the generated energy is used to recharge EVs. In case of a blackout, the EV batteries will serve as backup generators for the building (Colthorpe, 2014).

11.3.2.6 Issues and Improvements

Japan imports a large portion of the energy consumed and mainly uses fossil fuels. Therefore, although BEVs have no emissions, they are still being fueled primarily using fossil fuel generated electricity. Thus, Japan overall has not reduced carbon emissions nor other GHG emissions with the transition to EVs. In order for EVs to be environmentally friendly, Japan should source the majority of its electricity from clean energy, especially solar. Also, EVs should be charged using SPCSs to fuel the vehicles using clean energy and to reduce the strain on the electrical grid. Tesla and several other companies are working to install SPCSs, demonstrating that this technology is feasible for Japan to implement.

11.4 Australia

11.4.1 Top EV Car Sales

In Australia in the first quarter of 2015, the Mitsubishi Outlander PHEV had the highest sales followed closely by Tesla with the BMW i3 farther behind. Nissan Leaf and Holden Volt sales have significantly dropped from 2014 sales (EV Sales, 2015).

11.4.2 Current Status

In 2013, Australia's electricity generation was mainly coal, followed by natural gas, hydroelectricity, and some renewables. However, over the past two years there has been a push for increased renewable energy generation, especially solar (Australia, 2014).

The EV market is not on par with other major countries. In 2014, only about 1180 BEVs and PHEVs were sold total. There are few government subsidies to make the price of EVs more competitive with ICE vehicles in Australia. In addition, the charging station infrastructure is currently very limited in Australia and is a barrier to EV adoption. However, the greatest inhibitor of EV adoption is the fact that charging an EV in Australia results in more carbon dioxide emissions than burning liquefied fossil fuels in an ICE vehicle. Consumers do not feel as environmentally friendly driving an EV as consumers who live in countries with cleaner electricity generation; thus losing one of the greatest appeals of EVs (Duff, 2015).

The Central Area Transit (CAT) system of Perth conducted a cost benefit analysis and determined that the CAT bus fleet would have net savings of $6 million if the buses were all electric (Public transport, 2015). In addition, Energeia, which provides energy and electricity research advisory services to organizations, determined that the Australian economy would experience a net economic cost of $368 million if EVs were not adopted over the next 20 years. The economy would also lose the potential of an additional $878 million gross value added. In order to achieve full economic potential, Australia should have 4 million EVs on the road by 2035.

11.4.3 Policy

In 2009, at the United Nations Framework Convention on Climate Change (UNFCCC) in Copenhagen, Australia pledged to reduce carbon emissions by 5% or 15–25% (under different conditions of a global agreement that stabilizes GHG levels) relative to 2000 emission levels. Australia and other leading developed countries met in June 2015 and unofficially agreed to target a 40–70% reduction in carbon emissions compared to 2010 levels by 2050 (Evans and Yeo, 2015). However, countries' official carbon emission reduction pledges will be made at the UNFCCC in Paris from November 30–December 11, 2015 (Quantified economy-wide, 2014). Farmers, environmental groups, and organizations in Australia are urging the government to pledge these targets or even a zero carbon emissions target by 2050 (Australian Government, 2015).

The Australia Electric Vehicle Association Inc. (AEVA) is a nonprofit organization of individuals and organizations with an interest in EVs. The group works to educate consumers about EVs and promote EV adoption, as well as foster EV research and provide a forum for communication about EVs (Policies, 2013).

Green Vehicle Guide is an online service by the government to provide ratings on new AU vehicles based on GHG and air pollution emissions, calculated comparing vehicle test data to AU standards. Overall ratings are assigned using 0–5 stars and a greenhouse and air pollution rating, which ranges from 0–10 (best). This Guide enables consumers to compare all vehicle makes and models for up to three auto manufacturers to understand the vehicles' benefits and economic savings (Green Vehicle Guide, 2015).

Through the National Energy Efficiency Initiative (NEEI), the Australia Government granted $100 million to *Smart Grid, Smart City,* led by Ausgrid, to develop a smart grid on a commercial scale. The program tested smart grid technologies, measuring the benefits and costs of nationwide development. Thirty thousand households participated in this program from 2010–2014. Also, as a part of the program Ausgrid added 20 electric cars to its fleet to study how the EVs affect the grid. The Final Report on these studies has not yet been released (About Smart Grid, Smart City, 2015).

11.4.4 Incentives

Fuel-efficient vehicles in Australia are exempt from the Federal Luxury Car Tax. Vehicles that are purchased which are above a threshold value are subject to a higher tax rate for the proportion that exceeds the threshold. Fuel-efficient vehicles have a higher threshold value of $75,375 in 2014–2015 compared to $60,136 for other vehicles. The Emission Reduction Fund (ERF) offers incentives to consumers who reduce the emission intensity of Australia's transport sector. The ERF targets large fleets, such as rental companies or public bus fleets, and thus primarily promotes fleets, rather than individual consumers. However, this incentive has limited benefit for EVs since the majority is powered using the electrical grid, which has high carbon emissions. EVs qualify for a 20% reduction on their registration fees. The Green Vehicles Duty Scheme is another incentive for alternative fuel vehicles, which provides a stamp duty discount for vehicles based on GHG and air pollution ratings. Vehicles with A-rating, which includes EVs, are exempt from stamp duty (Review of alternative fuel, 2015).

11.4.5 Infrastructure

There are limited publicly accessible charge stations currently in Australia. However, there are efforts to increase the infrastructure. Beginning in 2015, the Royal Automobile Club (RAC) is developing an electric highway of public electric vehicle fast charge stations between Perth and South West. The goal is to provide a charging network to stimulate greater EV demand in Western Australia. Thirteen charge stations with two charging points each will be installed using E-station charge stations (RAC electric highway, 2015). E-station is an EVSE provider in Australia that sells charge stations for the home, office, car parking lot, or street parking. The charge stations are compatible for electric cars or buses (Welcome to E-station, 2015).

Similarly, Queensland also wants to build a 1600-km long electric highway of fast charging charge stations, many of which will be SPCSs, along the Bruce Highway. The first charge station of the electric highway will be an

SPCS, the first in Australia, and will be installed in Townsville. In addition, to subsidize costs, the solar provider Ergon Energy is offering to lease 25 kW of solar panels for businesses, and the Economic Development Queensland is offering to lease EV charging equipment (Queensland plans, 2015).

Tesla has plans to build a Supercharger network in Australia and power it with solar. Sixteen Superchargers will be built, connecting Melbourne with Brisbane and the Sunshine Coast by 2016. Goulburn Visitor's Centre was chosen as the first installation location. Tesla is also establishing hotel partnerships to install Level 1 and Level 3 wall-mounted charging units (Hall, 2015).

11.4.6 Issues and Improvements

Incentives should be provided for the purchase of EVs to make them more appealing to consumers. Removing the import barrier on foreign vehicles could reduce the premium currently paid for the vehicles. In addition, introducing a pricing scheme that charges consumers more for vehicles with higher GHG emissions would reduce the price gap between EVs and ICE vehicles and could encourage more consumers to purchase EVs. Incentives beyond subsidies, such as granting EVs priority access to bus lanes and providing designated parking spots, are also helpful for encouraging EV adoption.

There are four models of BEVs and 20 HEV models available in Australia. Providing consumers a greater selection of clean vehicles could assist them in finding the best vehicle for their needs and increase the likelihood of purchasing an EV. In addition, Australian consumers have cited the lack of charge station infrastructure as a barrier to EV adoption. Efforts to build charge station infrastructure should continue to increase. SPCSs should definitely be a focus of Australia's charge station infrastructure since the fossil fuel powered electrical grid discourages consumers from EV purchases. SPCSs use clean solar energy to power EVs, reducing emissions and the vehicles' overall impact on the environment.

11.5 Conclusions

This review of progress in many countries indicates that charging infrastructure and incentives are important for widespread EV adoption. High prices for gasoline and carbon dioxide emission taxes encourage EV purchases. In light of the December 2015 Paris Agreement, countries likely will be further encouraged to reduce carbon emissions through EVs.

The majority of countries reviewed in this chapter have relatively successful EV policies and initiatives, and they can share and learn from each other. It is crucial that countries engage in meaningful discussions with one another to aid in the global transition to EVs. National standards for EVs and EVSEs are important to achieve both for internal consistency and to eventually have worldwide standards. The political and technological barriers inhibiting widespread EV adoption must be removed. SPCSs must be sufficiently prevalent worldwide to combat both range anxiety and the emissions associated with the electrical grid.

References

About E.ON e-Mobility. 2015. E.ON e-Mobility. Accessed August 8, 2015. http://www.eon.dk/e-mobility/Om-EON-e-Mobility/.

About Smart Grid, Smart City. 2015. Smart Grid, Smart City. Accessed August 6, 2015. http://www.smartgridsmartcity.com.au/About-Smart-Grid-Smart-City.aspx.

Addressing climate change. 2015. IEA—International Energy Agency. Accessed August 6, 2015. http://www.iea.org/policiesandmeasures/climatechange/?country=Denmark.

Advansolar. Advansolar. Accessed August 4, 2015. http://www.advansolar.com/en/.

Akcayoz De Neve, P. 2015. "Electric vehicles in China." Belfer Center for Science and International Affairs. Accessed August 7, 2015. http://belfercenter.ksg.harvard.edu/publication/24345/electric_vehicles_in_china.html.

Amsterdam electric. 2015. Amsterdam.nl. Accessed August 7, 2015. https://www.amsterdam.nl/parkeren-verkeer/amsterdam-elektrisch/.

Australia. 2014. U.S. Energy Information Administration (EIA). http://www.eia.gov/beta/international/analysis.cfm?iso=AUS.

Australian Government urged to adopt a zero carbon emissions target by 2050. 2015. *The Guardian*. Accessed August 6, 2015. http://www.theguardian.com/environment/2015/jun/16/australian-government-urged-to-adopt-a-zero-carbon-emissions-target-by-2050.

Autolib'. 2015. Autolib' Paris Website. Accessed August 5, 2015. https://www.autolib.eu/en/.

Ayre, J. 2015a. "France offering up to €10,000 to switch from old diesel cars to electric cars." *CleanTechnica*. http://cleantechnica.com/2015/02/16/france-offering-drivers-old-diesel-cars-e10000-switch-plug-ins/.

Ayre, J. 2015b. "Japan now home to more electric vehicle charging stations than gas stations." *CleanTechnica*. http://cleantechnica.com/2015/02/18/japan-now-home-electric-vehicle-charging-stations-gas-stations/.

Bae, H. and D. Hurst. 2012. *Electric two-wheel vehicles in Asia-Pacific*. Boulder, CO: Pike Research. Accessed August 5, 2015. http://www.navigantresearch.com/wp-content/uploads/2012/04/ETVAP-12-Executive-Summary.pdf.

Block, B. 2015. "Germany boosts electric vehicle development." Worldwatch Institute. http://www.worldwatch.org/node/6251.

CHAdeMO. 2015. CHAdeMO Association. Accessed August 4, 2015. http://www.chademo.com/.

China. 2015. U.S. Energy Information Administration (EIA). http://www.eia.gov/beta/international/analysis.cfm?iso=CHN.

China electric car sales—June. 2015. *EVObsession*. Accessed July 17, 2015. evobsession.com/byd-qin-still-crushing-it-in-china-china-electric-car-sales-june/.

China electric vehicle charging station and charging pile report 2015–2016. 2015. China Electricity Market. Accessed August 5, 2015. http://www.researchandmarkets.com/research/ccbd9p/china_electric.

China top sector E-mobility. 2014. 1–23. Web. Dec. 15, 2015. http://china.nlambassade.org/binaries/content/assets/postenweb/c/china/zaken-doen-in-china/high-tech/china-top-sector-e-mobility-opportunity-report.pdf.

China urges local governments to buy more new-energy cars. 2015. Reuters. Last modified July 13, 2014. http://www.reuters.com/article/2014/07/14/us-china-electriccar-idUSKBN0FJ08Y20140714.

CLEVER. 2015. CLEVER. Accessed August 5, 2015. https://www.clever.dk/english/.

Climate action tracker. 2015. Climate Action Tracker. http://climateactiontracker.org/countries.html.

Climate case. 2015. Urgenda. http://www.urgenda.nl/en/climate-case/.

Colthorpe, A. 2014. "Kyocera installs 230kW solar charging station with EV battery backup." *PV-Tech*. http://www.pvtech.org/news/kyocera_installs_230kw_solar_charging_station_with_ev_battery_backup.

Countries. IEA—International Energy Agency. 2015. Accessed August 6, 2015. http://www.iea.org/countries/.

Cremer, A., A. Williams, and T. Severin. 2014. "Germany plans electric car motorway charging stations." Reuters UK. http://uk.reuters.com/article/2014/12/27/germany-electric-cars-idUKL6N0UB0AJ20141227.

DeMorro, C. 2015a. "China unleashing another beast of an electric car incentive." *CleanTechnica*. http://cleantechnica.com/2015/06/05/china-unleashing-another-beast-of-an-electric-car-incentive/.

DeMorro, C. 2015b. "2016 BYD E6 will get bigger battery & longer range." *CleanTechnica*. n.p., Web. Dec. 15, 2015. http://cleantechnica.com/2015/05/21/2016-byd-e6-will-get-bigger-battery-longer-range/.

Detailed presentation of the 14 actions. 2012. Ministère Du Développement Durable. http://www.developpement-durable.gouv.fr/Presentation-detaillee-des-14,26839.html.

Duff, C. 2015. "Why Australians aren't buying electric cars." *CarsGuide*. http://www.carsguide.com.au/car-news/why-australians-arent-buying-electric-cars-yet-30869#.VcLS6ZNViko.

Eco2charge: A project designed to accelerate the deployment of electric vehicles. 2014. Accessed August 3, 2015. http://www.eco2charge.fr/2014-09-18%20CP%20final%20Eco2charge_GB.pdf.

Ecotricity. 2015. Ecotricity. Accessed July 28, 2015. http://www.ecotricity.co.uk/.

ElaadNL. 2015. ElaadNL. Accessed July 31, 2015. http://www.elaad.nl/.

Electricity generation. 2015. Energy UK. Accessed August 10, 2015. http://www.energy-uk.org.uk/energy-industry/electricity-generation.html.

Energy concept. 2015. Accessed August 5, 2015. http://www.iea.org/policiesandmeasures/pams/germany/name-34991-en.php.

Energy: Energy transition to green growth. 2015. Assemblée Nationale. Accessed August 4, 2015. http://www.assemblee-nationale.fr/14/dossiers/transition_energetique_croissance_verte.asp.

EU to support development of electric vehicle transport roads in northern Europe. 2015. Accessed February 9, 2015. http://ec.europa.eu/inea/en/ten-t.

EVAAP. 2015. EVAAP. Accessed August 5, 2015. http://www.evaap.org/info/info.html?sgubun=1.

Evans, S. and S. Yeo. 2015. "G7 leaders target zero-carbon economy." *Carbon Brief.* http://www.carbonbrief.org/blog/2015/06/g7-leaders-target-zero-carbon-economy/.

EV Norway. 2015. Accessed August 7, 2015. http://www.evnorway.no/.

EVolution electric vehicles in Europe: Gearing up for a new phase? 2014. Accessed August 6, 2015. http://www.mckinsey.com/~/media/McKinsey%20Offices/Netherlands/Latest%20thinking/PDFs/Electric-Vehicle-Report-EN_AS%20FINAL.ashx.

EV Sales. 2015. EV Sales. http://ev-sales.blogspot.com/search/label/Europe.

Fastned—Home. 2015. Fastned. Accessed August 7, 2015. http://fastned.nl/nl/.

Gao, P., C. Malorny, S. Mingyu Guan, T. Wu, T. Luk, L. Yang, D. Lin, and X. Xu. April 2015. "Supercharging the development of electric vehicles in China." *McKinsey China*. McKinsey. Web. Dec. 20, 2015. http://www.mckinseychina.com/wp-content/uploads/2015/04/McKinsey-China_Electric-Vehicle-Report_April-2015-EN.pdf?5c8e08.

Global EV outlook 2015 (Infographic). 2014. Retrieved from http://www.iea.org/evi/Global-EV-Outlook-2015-Update_1page.pdf.

Gordon-Bloomfield, N. 2014. "Vive la loiture electrique! Normandy spurs electric car revolution with massive discounts." *Transport Evolved*. https://transportevolved.com/2014/04/14/vive-la-voiture-electrique-normandy-spurs-electric-car-revolution-massive-discounts/.

Government initiative for promoting next generation vehicles. 2014. Accessed August 7, 2015. http://mddb.apec.org/Documents/2014/AD/AD1/14_ad1_025.pdf.

Green Motion. 2015. Green Motion. Accessed August 7, 2015. http://greenmotion.com/.

Green Vehicle Guide. 2015. An Australian Government Initiative. Accessed August 7, 2015. https://www.greenvehicleguide.gov.au/GVGPublicUI/Home.aspx.

Hall, S. 2015. Electric car company Tesla plans 16 Supercharger stations between Melbourne and Brisbane. *Drive*. Accessed August 7, 2015. http://www.drive.com.au/it-pro/electric-car-company-tesla-plans-16-supercharger-stations-between-melbourne-and-brisbane-20150111-12lw2a.

Heymann, E. 2014. "Co2 emissions from cars." n.p.: Deutsche Bank Research. Accessed July 27, 2015. https://www.dbresearch.com/.

Japan. 2015. U.S. Energy Information Administration (EIA). http://www.eia.gov/beta/international/analysis.cfm?iso=JPN.

Japan automakers advance electric charging infrastructure with new company, Nippon Charge Service. 2014. Nissan Motor Company Global Website. http://www.nissan-global.com/EN/NEWS/2014/_STORY/140530-01-e.html.

Kane, M. 2015. "UK & Netherlands account for two thirds of PHEV sales in Europe." *Inside EVs*. http://insideevs.com/uk-netherlands-account-for-two-thirds-of-phev-sales-in-europe/.

Krieger, A., P. Radtke, and L. Wang. 2012. "Recharging China's electric-vehicle aspirations." McKinsey & Company. n.p. Web. Dec. 15, 2015. http://www.mckinsey.com/insights/energy_resources_materials/recharging_chinas_electric-vehicle_aspirations.

Le Groupe La Poste. 2015. Le Groupe La Poste. Accessed August 3, 2015. http://legroupe.laposte.fr/.

Loveday, E. 2015a. "Battery swapping a reality in China." *Inside EVs*. http://insideevs.com/china-battery-swap/.
Loveday, E. 2015b. "BYD electric bus sales." *Inside EVs*. http://insideevs.com/byd-electric-bus-sales-4000-per-year-china/.
Making the connection: The plug-in vehicle infrastructure strategy. 2011. London: Department of Transport. Accessed August 2, 2015. https://www.gov.uk/government/organisations/department-for-transport.
Market and operations. 2015. Statnett—Fremtiden Er Elektrisk. Accessed July 24, 2015. http://www.statnett.no/en/Market-and-operations/.
Marro, N., H. Liu, and Y. Yan. 2015. "Opportunities and challenges in China's electric vehicle market." *China Business Review*. http://www.chinabusinessreview.com/opportunities-and-challenges-in-chinas-electric-vehicle-market/.
Masson, L.J. 2013. "German electric vehicle players launch charging info-exchange network." *PluginCars*. http://www.plugincars.com/intercharge-hubject-european-network-charging-stations-127430.html.
Mission. 2015. *Dansk Elbil Alliance*. n.p. Web. Dec. 15, 2015. http://www.danskelbilalliance.dk/English.aspx.
Mock, P. and Z. Yang. *Driving electrification*. Washington, DC: n.p., 2014. Accessed August 4, 2015. http://www.theicct.org/sites/default/files/publications/ICCT_EV-fiscal-incentives_20140506.pdf.
Morris, C. 2015. "Uber adds PHEVs to its fleets in China." *Charged Electric Vehicles Magazine*. https://chargedevs.com/newswire/uber-adds-phevs-to-its-fleet-in-china/.
Nagatsuka, S. 2014. "The world auto industry: Situation and trends." Japan Automobile Manufacturers Association. http://www.oica.net/wp-content/uploads/JAMA_-Situation-and-trends.pdf.
Nanjing Public Transportation Group CO. LTD. 2015. Nanjing Public Transportation Group CO. LTD. http://www.njgongjiao.com/.
National electromobility development plan. 2015. Germany Trade and Invest. Accessed August 4, 2015. http://www.gtai.de/GTAI/Navigation/EN/Invest/Industries/Smarter-business/Smart-mobility/national-electromobility-development-plan.html#384066.
Open charge alliance. 2015. Open Charge Alliance. Accessed July 26, 2015. http://www.openchargealliance.org/?q=node/10.
Our best practices. 2015. Next generation vehicle promotion center. Accessed August 5, 2015. http://www.cev-pc.or.jp/english/practice/index.html.
Overview for purchase and tax incentives for EVs in the EU in 2015. 2015. ACEA—European Automobile Manufacturers' Association. http://www.acea.be/uploads/publications/Electric_vehicles_overview_2015.pdf.
Overview of E+ Partners. 2015. Electromobility+. Accessed August 4, 2015. http://electromobility-plus.eu/?page_id=769.
Policies. 2013. The Australian Electric Vehicle Association. http://www.aeva.asn.au/wiki/policies.
Public transport: The University of Western Australia. 2015. The University of Western Australia. Accessed August 7, 2015. http://www.uwa.edu.au/university/transport/public-transport.
Quantified economy-wide emission reduction targets by developed country Parties to the Convention. 2014. n.p.: United Nations. Accessed August 6, 2015. http://unfccc.int/resource/docs/2014/tp/08.pdf.

Queensland plans 1,600 km string of fast-charging stations for electric cars. 2015. *The Guardian*. Accessed August 7, 2015. http://www.theguardian.com/australia-news/2015/jul/25/queensland-to-encourage-fast-charging-stations-to-service-electric-cars-statewide.

RAC electric highway. 2015. RAC WA. Accessed August 7, 2015. http://rac.com.au/news-community/environment/electric-highway-and-electric-vehicles.

Sett. 390 S (2011–2012). 2012. Energy and the Environment Committee. https://www.stortinget.no/no/Saker-og-publikasjoner/Publikasjoner/Innstillinger/Stortinget/2011-2012/inns-201112-390/?lvl=0#a2.8.

Review of alternative fuel vehicle policy targets and settings for Australia. July 2015. Accessed August 4, 2015. http://www.esaa.com.au/Library/PageContentFiles/69ae0935-d7e1-4dfe-9d3d-0309a1ff8e62/Energeia%20Report%20for%20esaa%20_%20Optimal%20AFV%20Policy%20Targets%20and%20Settings%20for%20Australia.pdf.

RWE-Mobility. RWE Group. 2015. Accessed August 5, 2015. http://www.rwe.com/web/cms/en/183210/rwe/innovation/projects-technologies/energy-application/e-mobility/.

SGCC decontrols construction of charging. 2014. Hanergy. http://www.hanergy.com/en/content/details_36_1349.html.

Shahan, Z. 2014. "Electric cars 2015." *EV Obsession*. n.p. Web. Dec. 15, 2015. http://evobsession.com/electric-cars-2014-list/.

Sharman, A. 2015. "Power struggle stalls London's electric cars." *Financial Times*. http://www.ft.com/home/us.

Solar charging station. 2010. Berlin, Germany: n.p. Accessed August 6, 2015. http://www.younicos.com/download/Yana/solar_charging_mobility_Adlershof_en.pdf.

Strategic roadmap for plug-in electric and hybrid vehicle charging infrastructure. 2009. Accessed July 31, 2015. http://www.ademe.fr/sites/default/files/assets/documents/88761_roadmap-plug-in-electric-and-hybrid-vehicle-charging-infra.pdf.

Tan, Q., M. Wang, Y. Deng Rao, and X. Zhang. 2014. "The cultivation of electric vehicles market in China: Dilemma and solution." MDPI. Web. Dec. 14, 2015. http://www.mdpi.com/2071-1050/6/8/5493/pdf.

Tesla Motors. 2015a. Hanergy. Accessed August 6, 2015. http://hanergy.eu/teslamotors/.

Tesla Motors. 2015b. Tesla Motors | Premium Electric Vehicles. Accessed August 6, 2015. http://www.teslamotors.com/.

Tesla Motors and Athlon Car Lease announce electric vehicle leasing program in Europe. Last modified July 2011. Accessed August 6, 2015. http://www.teslamotors.com/blog.

Tesla to Invest in Charging Infrastructure in Japan-Nikkei. CNBC. Last modified April 29, 2015. http://www.cnbc.com/2015/04/29/reuters-america-tesla-to-invest-in-charging-infrastructure-in-japan-nikkei.html.

Tillemann, L. "China's electric car boom: Should Tesla Motors worry?" *Fortune*. Last modified February 19, 2015. http://fortune.com/2015/02/19/chinas-electric-car-boom-should-tesla-motors-worry/.

Timmons, H. "Consider the E-bike: Can 200 million Chinese be wrong?" *Quartz*. n.p., Oct. 22, 2013. Web. Dec. 15, 2015. http://qz.com/137518/consumers-the-world-over-love-electric-bikes-so-why-do-us-lawmakers-hate-them/.

Toshiba Corporation to start pilot project of microelectronic vehicle in Miyako Island. Toshiba. Last modified February 21, 2013. http://www.toshiba.co.jp/about/press/2013_02/pr2102.htm.

Tost, D. 2014a. "Berlin approves new incentives for electric car drivers." EurActiv. http://www.euractiv.com/sections/transport/berlin-approves-new-incentives-electric-car-drivers-308700.

Tost, D. 2014b. "Germany to miss target for one million e-cars by 2020." EurActiv. http://www.euractiv.com/sections/energy/germany-miss-target-one-million-e-cars-2020-310523.

Vergis, S., T.S. Turrentine, L. Fulton, and E. Fulton. Plug-in electric vehicles: A case study of seven markets. October 2014. Accessed August 4, 2015. http://www.its.ucdavis.edu/research/publications/.

Welcome to E-station. 2015. E-Station. Accessed August 7, 2015. http://e-station.com.au/.

World premiere Utrecht charging station all electric cars. Utrecht. Last modified June 9, 2015. http://utrecht.nieuws.nl/stad/39019/wereldprimeur-in-utrecht-laadpaal-voor-alle-elektrische-autos/.

Yu, R. "Tesla Cars to Meet China Charging Standards." Tesla Motors. Last modified April 12, 2015. http://my.teslamotors.com/it_CH/forum/forums/tesla-cars-meet-china-charging-standards.

Zach. 2015, July 24. Europe vs North America vs China EV sales. Retrieved August 4, 2015, from http://evobsession.com/europe-vs-north-america-vs-china-ev-sales-charts/.

ZAE signed the deal with China's Bustil to design 7,000 swap stations for Nanjing. 2015. ZIV-AV Engineering L.T.D. Accessed August 7, 2015. http://www.zivaveng.com/rec/430-ZAE-signed-the-deal-with-Chinas-Bustil-t.

2010 to 2015 government policy: Transport emissions. Last modified May 8, 2015. https://www.gov.uk/government/organisations/department-for-transport.

12

Conclusions

Larry E. Erickson, Gary Brase, and Jackson Cutsor

CONTENTS

Change is the law of life. And those who look only to the past or present are certain to miss the future.

John F. Kennedy

12.1 Summary of Progress

In the last several years there has been progress with respect to many of the topics in this book. Electric vehicles (EVs) continue to be sold in significant quantities. Tesla delivered more than 50,000 new cars in 2015, and their infrastructure of Supercharger stations continues to grow (Waters, 2016). Tesla provides a good example of the importance of an infrastructure of solar powered charging stations (SPCSs). The prices of solar panels and batteries for EVs and energy storage have continued to fall over the past several years. A significant fraction of new electric generating capacity is powered by either wind or solar energy. The path forward leads to a world in which EVs will be the vehicle of choice and solar and wind energy will be less expensive than other alternatives. Battery storage of energy is becoming competitive, and it is being used with smart grid developments and time-of-use prices to deliver electric power efficiently and effectively.

12.2 Important Research and Development Challenges

Research to develop improved solar panels and batteries continues to have great value. Better efficiency in converting solar energy to electricity is expected in the next 25 years. Further reduction in cost is expected as well. Progress in batteries to increase energy density, lengthen battery life, and reduce cost will have value for EVs and for energy storage. While there is a continuing growth in new brands and models of EVs, research and development to optimize EVs is continuing on many fronts. Similar further research and development is needed to improve SPCSs.

Decision support systems are needed to manage the smart grid through time-of-use (TOU) prices and automated systems that reduce peak power requirements and balance supply and demand effectively. When more than 50% of vehicles are EVs and more than 50% of electricity is generated by wind and solar energy, new decision support software will be needed to manage the smart grid efficiently. Research to improve decision support systems for grids with large numbers of SPCSs, EVs, and energy storage is needed. The educational materials for customers regarding their interaction with smart grid communication systems and decision support systems with TOU prices and automation features need further study and development. It is known that high temperatures may impact the lifetime of batteries. Research is needed to understand the relative importance of shade provided by SPCSs in very hot parking lot environments such as Arizona. Temperature measurements can be made with and without solar panels providing shade and data on battery life can be collected in environments with high summer temperatures.

Research and development to make wireless EV charging easy, efficient, and inexpensive would have great value. The cord to connect to the vehicle is of greater concern when it is used in public SPCSs because of the need for safety. A worldwide standard for EV cords is necessary. On the other hand, eliminating the need for a cord also has significant value.

12.3 Integration and Implementation of New Developments

Several new developments have the potential to be very important in the effort to develop an infrastructure of SPCSs. The smart grid with time-of-use prices increases the value of energy generated with solar panels. Battery storage with SPCSs enables solar energy to be made available at times different from when it is generated. The smart grid with improved communication and time-of-use prices enables users to shift their energy use to more favorable times.

As the growth of solar energy generation systems fills parking lots and roof tops with solar panels, the infrastructure for EV charging will be more robust and supportive of EV transportation. If convenient charging is available at most locations where EVs are parked, the battery size needed for effective daily use is reduced.

The integration of SPCSs into the grid has significant value for society. Energy storage in batteries as part of a smart grid with TOU prices may help the management of supply and demand in grids with large numbers of SPCSs. There is significant demand for electricity near most SPCSs because parking lots usually have many other activities nearby. Policy issues related to the integration of electricity generated by SPCSs into the grid need to be managed appropriately such that benefits are distributed equitably and there are no barriers that prevent integration. Demand charges, TOU prices, and other smart grid features should be beneficial in the integration of SPCSs into the grid.

The SPCS system brings electricity to parking lots in a new way. This allows electronic advertising to be integrated into SPCSs at a reasonable cost, and it provides another alternative to pay for SPCS equipment and electricity consumed. Free Level 1 and Level 2 charging of EVs is inexpensive and it is being provided in some parking lots with costs paid for by advertising, an employer, or retail establishments. New business models can be developed that integrate SPCSs into the many common parking lot business models that are in use. Because electric power is generated with SPCSs, there is the business model of the electric utility owning or leasing the SPCSs and producing power for its grid, including selling to EV customers that use the SPCSs (Robinson et al., 2014). Where electric utilities are regulated, they may need to obtain approval to sell electricity to customers with EVs at SPCSs. The regulatory body would be involved in approving the rate, which might be a TOU rate.

Another integration topic for urban communities is air quality. An air quality initiative may include an effort to generate electricity with SPCSs and increase the infrastructure for EV charging in order to increase EV use and thereby improve urban air quality. There are many urban areas that have poor air quality such that it impacts health and the quality of life. The justification for installing SPCSs in urban parking lots may include the goal of improving air quality. The cost of free charging with SPCSs could be paid for with a sales tax or paid through utility billings for electricity, including EVs using the SPCSs.

12.4 Education and Achieving this Great Transition

Education is needed to accomplish the goals of electrifying transportation, adding an infrastructure of SPCSs, introducing smart grids with demand charges, TOU prices, decision support systems, and new communication features that encourage customers to optimize their electricity use. This transition to a more participatory system has many features that need to be understood sufficiently to work effectively. Peak power demands can be reduced when customers shift their use of electricity to another time. One of the purposes of this book is to begin to inform people of the opportunities and benefits of EVs, SPCSs, and the smart grid.

In addition to basic education about the nature of EVs, SPCSs, and the smart grid, there needs to be some consideration of how people react to these changes as they disrupt the traditional ICE market, petroleum-based energy, and the basic electrical grid most people have now. Change and disruption can be either upsetting or exciting, depending on how it is presented and who is experiencing it. Take, for example, the decision to purchase an EV. Several of the preceding chapters have explored the basic economics (including full cycle analyses) of having an EV versus an ICE. We are now at a point where an EV can be quite economically competitive with an ICE. But what influences when a particular individual decides to transition to an EV?

Surprisingly, some important factors are decidedly noneconomic; people who are made conscious of social status, and given the opportunity to buy a more expensive but environmentally "greener" product (a car, an appliance, or a cleaning product), are more likely to buy that greener product (Griskevicius, Tybur, and Van den Bergh, 2010). Similarly, people pay attention to their neighbors and feel more compelled to adopt new technologies such as home solar panels when their neighbors have done so (Plumer, 2015). This seems to come down to wanting to be seen by others as holding valued social traits (e.g., caring about the environment, having the ability to obtain the latest technologies).

This "keeping up with the Jones's" effect is not always a guarantee, however. Several studies have looked at what happens when people are given meters in their homes that show them how much electricity they are using, relative to the usage of others. These "smart meters" (as part of a smart grid) can lead individuals to decrease their own energy consumption, but effectiveness of these meters depends tremendously on how they are designed and how they are used (Ehrhardt-Martinez et al., 2010; Mooney, 2015).

Perhaps a larger concern is the array of forces opposed to education, or eventual adoption, of renewable energy and electric vehicles. Some of these forces are predictable, but others are not as obvious. Most of them share a common thread: as Upton Sinclair said, "It is difficult to get a man to understand something, when his salary depends upon his not understanding it." The petroleum industry consistently works against the adoption of alternative energy sources (Edelstein, 2014). Energy companies are often resistant to change, both because of their investments in oil and coal and because of issues related to the need to maintain/upgrade the electrical infrastructure (Bushnell, 2015). Tesla, which has been selling its electric cars directly to customers, has been frustrated by state laws (enacted with the support of dealerships) that limit or forbid car sales except through franchised dealerships (Chapman, 2013). Even dealerships selling their own EVs appear to be suppressing the sales of those vehicles over ICE models (Murphy, 2014; one reason appears to be the much higher maintenance and repair revenue dealerships get from ICE models).

Last, there are growing political dimensions to alternative energy and EV adoption. For instance, the Kansas state legislature in 2015 faced an almost $600 million budget deficit, primarily due to conservative-led slashing of state income taxes. Efforts to fix this shortfall included a variety of steps, but legislators were loath to raise taxes as a solution. With one exception; an early move to help with the deficit was to eliminate several incentives for renewable energy development. Curiously, "conservatives in the Legislature who have traditionally opposed tax increases have said renewable energy is one sector they don't mind turning to for revenue" (Himmelberg, 2015). This political dimension is actually a bit odd because the transition to renewable energy and EVs is key to fundamental issues across the political spectrum. Yes, renewable energy is good for the environment (a progressive goal), but it also is our best shot at independence from foreign oil and future economic prosperity in this area (a conservative goal).

12.5 The United Nations Paris Agreement on Climate Change

One of the important reasons for developing an infrastructure of SPCSs is to help meet the goals of the Paris Agreement to prevent global temperatures from increasing more than 1.5°C. It may be possible to generate at least 1/3 of the electricity used in the world by covering parking lots with SPCSs. Many of the concepts of adding SPCSs in parking lots can be implemented almost anywhere in the world. There are off-grid locations where SPCSs with inexpensive batteries can be used to provide electricity for lighting and other purposes as well. With good policies and efforts to eliminate any barriers, there is the potential to transition to renewable energy, SPCSs, EVs, and reduced greenhouse gas emissions. This can help accomplish the goals of the Paris Agreement (UNFCCC, 2015).

> You must be the change you wish to see in the world.
>
> **Mahatma Gandhi**

References

Bushnell, J. 2015. How (and who) will pay for our energy infrastructure? https://energyathaas.wordpress.com/2015/02/10/how-and-who-will-pay-for-our-energy-infrastructure/.

Chapman, S. 2013. Car buyers get hijacked. *Chicago Tribune*. Retrieved April 12, 2015. http://articles.chicagotribune.com/2013-06-20/news/ct-oped-0620-chapman-20130620_1_tesla-motors-car-dealers-car-costs.

Edelstein, S. 2014. Leaked playbook shows how big oil fights clean energy; http://news.discovery.com/tech/alternative-power-sources/leaked-playbook-shows-how-big-oil-fights-clean-energy-141204.htm.

Ehrhardt-Martinez, K., K.A. Donnelly, and J.A. Laitner. 2010. Advanced metering initiatives and residential feedback programs: A meta-review for household electricity-saving opportunities, Report Number E105; http://www.energycollection.us/Energy-Metering/Advanced-Metering-Initiatives.pdf.

Griskevicius, V., J. Tybur, and B. Van den Bergh. 2010. Going green to be seen: Status, reputation, and conspicuous conservation. *Journal of Personality and Social Psychology*, 98(3): 392–404.

Himmelberg, A. 2015. Kansas Legislature mulls slashing green energy incentives; http://ksnt.com/2015/03/21/kansas-legislature-mulls-slashing-green-energy-incentives/

Mooney, C. 2015. Why 50 million smart meters still haven't fixed America's energy habits; http://www.washingtonpost.com/news/energy-environment/wp/2015/01/29/americans-are-this-close-to-finally-understanding-their-electricity-bills/.

Murphy, T. 2014. Dealers blamed for dismal EV market; http://wardsauto.com/blog/dealers-blamed-dismal-ev-market.

Plumer, B. 2015. Solar power is contagious: Installing panels often means your neighbors will too; http://www.vox.com/2014/10/24/7059995/solar-power-is-contagious-neighbor-effects-panels-installation.

Robinson, J., G. Brase, W. Griswold, C. Jackson, and L. Erickson. 2014. Business models for solar powered charging stations to develop infrastructure for electric vehicles, *Sustainability* 6: 7358–7387.

UNFCCC. 2015. Paris Agreement, United Nations Framework Convention on Climate Change, FCCC/CP/2015/L.9, December 12, 2015; http://unfccc.int.

Waters, R. 2016. Tesla sales pace falls short at end of 2015. *Financial Times*, January 3, 2016; http://www.ft.com.

Index